AF478550

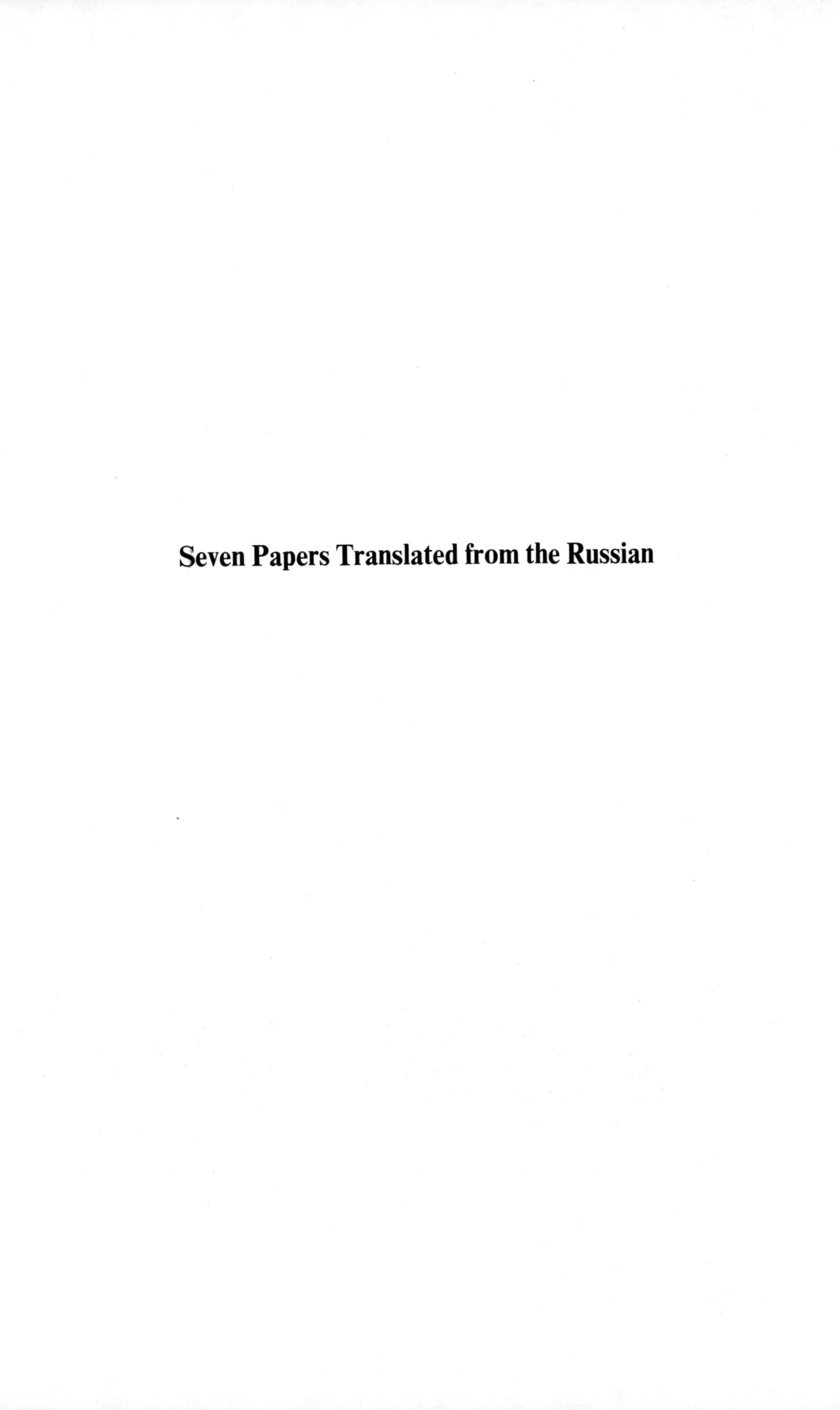

Seven Papers Translated from the Russian

**AMERICAN MATHEMATICAL SOCIETY
TRANSLATIONS**

Series 2

Volume 138

Seven Papers Translated from the Russian

by
I. V. Kovalishina and V. P. Potapov

AMERICAN MATHEMATICAL SOCIETY
PROVIDENCE, RHODE ISLAND

Translated by H. H. McFADEN
Translation edited by BEN SILVER

1980 *Mathematics Subject Classification* (1985 *Revision*). Primary 15Axx, 30D50, 30F05, 30G30, 47B50, 47H09; Secondard 47A99.

Library of Congress Cataloging-in-Publication Data

```
Kovalishina, I. V.
   [Selections. English.  1988]
   Seven papers translated from the Russian / by I.V. Kovalishina and
V.P. Potapov.
      p.   cm. -- (American Mathematical Society translations, ISSN
0065-9290 ; ser. 2, v. 138)
   Contents: The multiplicative structure of analytic real J
-expansive matrix-valued functions / I.V. Kovalishina and V.P.
Potapov -- General theorems on the structure and the splitting off
of elementary factors of analytic matrix-valued functions / V.P.
Potapov -- An indefinite metric in the Nevanlinna-Pick problem /
I.V. Kovalishina and V.P. Potapov -- Linear fractional
transformations of matrices / V.P. Potapov -- The radii of a Weyl
disk in the matrix Nevanlinna-Pick problem / I.V. Kovalishina and
V.P. Potapov -- A theorem on the modulus. I: Main concepts. The
modulus /V.P. Potapov -- II. [without special title] / V.P. Potapov.
   ISBN 0-8218-3114-3
   1. Mathematics--Collected works--Translations from Russian.
I. Potapov, V. P. (Vladimir Petrovich), 1914-1980.  II. Title.
III. Series.
QA3.A572 ser. 2, vol. 138
510 s--dc19
[510]                                                    87-33650
                                                             CIP
```

Contents

Russian Contents*

---

*The American Mathematical Society scheme for transliteration of Cyrillic may be found at the end of index issues of *Mathematical Reviews*.

Amer. Math. Soc. Transl.
(2) Vol. 138, 1988

The Multiplicative Structure of Analytic Real J-Expansive Matrix-Valued Functions

I. V. KOVALISHINA AND V. P. POTAPOV

$1°$. We study an analytic matrix-valued function $W(\lambda)$ satisfying the following two conditions:

1. $W(\lambda)JW^*(\lambda) - J \geq 0$ $(\operatorname{Re}\lambda > 0)$,
2. $\overline{W(\lambda)} = W(\bar{\lambda})$,

that is, a real matrix-valued function that is J-expansive in the right half-plane.

Obviously, a real matrix-valued function $W(\lambda)$ with a pole of multiplicity k at an imaginary point λ_0 has a pole of the same multiplicity at the conjugate point $\bar{\lambda}_0$, and the coefficients in the Laurent series expansions in neighborhoods of these points are complex conjugates.

The simplest real matrix-valued function $Z(\lambda)$ that is J-expansive in the right half-plane and J-unitary on the imaginary axis, has poles of first order at the points λ_0 and $\bar{\lambda}_0$, and is normalized by the condition $Z(\infty) = I$ is called an *elementary real factor* and written as follows:

$$Z(\lambda) = I + \frac{2\sigma_0 A}{\lambda - \lambda_0} + \frac{2\sigma_0 \overline{A}}{\lambda - \bar{\lambda}_0} \qquad (\sigma_0 = \operatorname{Re}\lambda_0).$$

Elementary factors will be considered for points λ_0 and $\bar{\lambda}_0$ lying in the right half-plane. For the case when λ_0 and $\bar{\lambda}_0$ $(\operatorname{Re}\lambda_0 > 0)$ are poles of $Z^{-1}(\lambda)$, the arguments in the proof are carried out similarly.

The function $Z(\lambda)$ can be decomposed into a product of two elementary factors with poles at the points λ_0 and $\bar{\lambda}_0$:

$$Z(\lambda) = \left(I + \frac{2\sigma_0 P_1}{\lambda - \lambda_0}\right)\left(I + \frac{2\sigma_0 P_2}{\lambda - \bar{\lambda}_0}\right) = b_{\lambda_0}(\lambda)b_{\bar{\lambda}_0}(\lambda).$$

Here $P_1^2 = P_1$, $P_1 J \to 0$, $P_2^2 = P_2$, and $P_2 J \geq 0$.

1980 *Mathematics Subject Classification* (1985 *Revision*). Primary 15A60, 30G30.

Translation of Izv. Akad. Nauk Armyan. SSR Ser. Fiz.-Mat. Nauk 18 (1965), no. 6, 3–10; MR **35** #5629.

The decomposition of $Z(\lambda)$ into a product of elementary factors can be realized in any order, and since $Z(\lambda) = \overline{Z(\bar{\lambda})}$, it follows that

$$Z(\lambda) = \left(I + \frac{2\sigma_0 P_1}{\lambda - \lambda_0}\right)\left(I + \frac{2\sigma_0 P_2}{\lambda - \bar{\lambda}_0}\right) = \left(I + \frac{2\sigma_0 \overline{P}_1}{\lambda - \bar{\lambda}_0}\right)\left(I + \frac{2\sigma_0 \overline{P}_2}{\lambda - \lambda_0}\right).$$

However, since multiplication does not commute, P_2 is not equal to $\overline{P}_1$ in general.

$2°$. Let us consider the expansion of a matrix-valued function $W(\lambda)$ in a Laurent series in a neighborhood of a point λ_0:

$$W(\lambda) = \frac{2\sigma_0 c}{(\lambda - \lambda_0)^k} + \cdots.$$

We choose any solution x of the equation $c^* J c x = c^*$ and form the matrix $P_1 = cxJ$.

It is not hard to see that P_1 does not depend on the choice of x and is a J-nonnegative projective matrix; that is, $P_1^2 = P_1$ and $P_1 J \geq 0$.

Further, the elementary factor $b_{\lambda_0}(\lambda) = (I + 2\sigma_0 P_1/(\lambda - \lambda_0))$ splits off from $W(\lambda)$, and the matrix-valued function $W_1(\lambda) = b_{\lambda_0}^{-1}(\lambda)W(\lambda)$ has the following properties: $W_1(\lambda)$ is J-expansive in the right half-plane, and the order of the pole of $W_1(\lambda)$ at the point λ_0 is reduced by 1.

Similarly, the elementary factor $b_{\bar{\lambda}_0} = (I + 2\sigma_0 P_2/(\lambda - \bar{\lambda}_0))$ splits off from $W_1(\lambda)$.

It can be proved that the product

$$Z(\lambda) = \left(I + \frac{2\sigma_0 P_1}{\lambda - \lambda_0}\right)\left(I + \frac{2\sigma_0 P_2}{\lambda - \bar{\lambda}_0}\right) = b_{\lambda_0}(\lambda) \cdot b_{\bar{\lambda}_0}(\lambda)$$

is an elementary real factor, and $b_{\lambda_0}^{-1}(\lambda)b_{\bar{\lambda}_0}^{-1}(\lambda)W(\lambda) = W_2(\lambda)$ remains a real matrix-valued function.

This splitting-off process leads to our main result:

THEOREM 1. *A real rational matrix-valued function $W(\lambda)$ that is J-expansive in the right half-plane and J-unitary on the imaginary axis decomposes into a product of real elementary factors:*

$$W(\lambda) = Z_1(\lambda) \cdot Z_2(\lambda) \cdots Z_k(\lambda)V,$$

where V is a constant real J-unitary matrix.

It follows from Theorem 1 that the study of a real matrix-valued function reduces to the study of the structure of an elementary real factor.

$3°$. The condition that $Z(\lambda)$ be real imposes the following restrictions on P_1 and P_2:

$$\overline{P}_1\left(I + \frac{i\sigma_0}{\tau_0}\overline{P}_2\right) = \left(I + \frac{i\sigma_0}{\tau_0}P_1\right)P_2, \tag{3.1}$$

which is a necessary and sufficient condition.

The question arises of whether there is a J-positive projection P_2 satisfying (3.1) for any J-positive projection P_1. We have proved

THEOREM 2. *The factor*

$$Z(\lambda) = \left(I + \frac{2\sigma_0 P_1}{\lambda - \lambda_0}\right)\left(I + \frac{2\sigma_0 P_2}{\lambda - \bar{\lambda}_0}\right)$$

is an elementary real factor, $Z(\lambda) = \overline{Z(\bar{\lambda})}$, if and only if

$$P_1\left(J - \frac{\sigma_0^2}{|\lambda_0|^2}\overline{P}_1 J\right)P_1^* \geq 0. \tag{3.2}$$

However, the construction of P_2 and P_1 leads to a fairly complicated structure for P_2. Therefore, it is more expedient to solve this problem for projections of rank 1, proving then that any elementary real factor decomposes into a product of finitely many factors of rank 1.

A real factor

$$Z(\lambda) = I + \frac{2\sigma_0 A}{\lambda - \lambda_0} + \frac{2\sigma_0 \overline{A}}{\lambda - \bar{\lambda}_0},$$

with A a matrix of rank 1 will be called a *real factor of rank* 1, or a *primary real factor.*

If a primary real factor is written as a product of two complex factors, then it is obvious that p_1 and p_2 are projections of rank 1; that is, $p_1 = Jg_1^* g_1$, $g_1 J g_1^* = 1$, $p_2 = J g_2^* g_2$, $g_2 J g_2^* = 1$, and

$$z(\lambda) = \left(I + \frac{2\sigma_0}{\lambda - \lambda_0}p_1\right)\left(I + \frac{2\sigma_0}{\lambda - \bar{\lambda}_0}p_2\right).$$

Note that the vector g is determined by the projection p to within a scalar factor e^{ix}.

Equality (3.1), written for projections of rank 1, gives us

THEOREM 3. *Suppose that $(I + 2\sigma_0 p_1/(\lambda - \lambda_0))$, where $p_1 = Jg_1^* g_1$, $g_1 J g_1^* = 1$, is an elementary factor of rank 1. Then there exists a factor*

$$\left(\frac{I + 2\sigma_0 p_2}{(\lambda - \bar{\lambda}_0)}\right)$$

of rank 1 supplementing the first factor to form a real factor

$$z(\lambda) = \left(I + \frac{2\sigma_0 p_1}{\lambda - \lambda_0}\right)\left(I + \frac{2\sigma_0 p_2}{\lambda - \bar{\lambda}_0}\right)$$

if and only if

$$|\bar{g}_1 J g_1^*| < \sec\theta \qquad (\theta = \arg\lambda_0).$$

Here the basis vectors g_1 and g_2 can be normalized so that

$$\bar{g}_1 J g_1^* = \tanh a \sec\theta e^{i\theta}, \tag{3.3}$$

and $g_2 = \cosh a\bar{g}_1 - \sinh a g_1.$

Since g_2 is determined by g_1 uniquely to within a scalar factor e^{ix}, the factor $(I + 2\sigma_0 p_2/(\lambda - \bar{\lambda}_0))$, and hence also the real factor of rank 1

$$z(\lambda) = \left(I + \frac{2\sigma_0 p_1}{\lambda - \lambda_0}\right)\left(I + \frac{2\sigma_0 p_2}{\lambda - \bar{\lambda}_0}\right)$$

$$= I + \frac{2\sigma_0 \cosh \alpha}{\lambda - \lambda_0} J g_1^* \bar{g}_2 + \frac{2\sigma_0 \cosh a}{\lambda - \bar{\lambda}_0}\overline{J g_1^* \bar{g}_2},$$

is uniquely determined by g_1. Therefore, we use the standard notation for the vectors g_1 and g_2:

$$g_1 = g, \qquad g_2 = \tilde{g} = \cosh a\bar{g} - \sinh ag.$$

In the subsequent presentation a vector $g_1 = g$, normalized by (3.3), will be called a *properly normalized vector*.

4°. In conclusion we prove that an elementary real factor decomposes into a product of primary real factors.

It is easy to see that if

$$Z(\lambda) = \left(I + \frac{2\sigma_0 P_1}{\lambda - \lambda_0}\right)\left(I + \frac{2\sigma_0 P_2}{\lambda - \bar{\lambda}_0}\right)$$

is an elementary real factor and if g_1 is any basis vector in the space $M_1 = HP_1$, then $|\bar{g}_1 J g_1^*| < \sec \theta$.

This circumstance enables us to properly normalize any basis vector $g_1 \in HP_1$ and to construct a vector $\tilde{g}_1 = \cosh a\bar{g}_1 - \sinh ag_1$, with the product

$$b_{\lambda_0}(\lambda)b_{\bar{\lambda}_0}(\lambda) = \left(I + \frac{2\sigma_0 J g_1^* g_1}{\lambda - \lambda_0}\right)\left(I + \frac{2\sigma_0 J \tilde{g}_1^* \tilde{g}_1}{\lambda - \bar{\lambda}_0}\right)$$

of the factors a primary real factor. However, it is not clear whether such a primary factor splits off from $Z(\lambda)$.

From the first complex elementary factor in $Z(\lambda)$ we split off the complex primary factor generated by the properly normalized basis vector $g_1 \in HP_1$, and from the second we split off the complex primary factor generated by the basis vector $g_2 \in HP_2$. Then

$$\left(I + \frac{2\sigma_0 P_1}{\lambda - \lambda_0}\right) = \left(I + \frac{2\sigma_0 p_1}{\lambda - \lambda_0}\right)\left(I + \frac{2\sigma_0 Q_1}{\lambda - \lambda_0}\right),$$

where $Q_1 = P_1 - p_1$, $p_1 = J g_1^* g_1$, and

$$\left(I + \frac{2\sigma_0 P_2}{\lambda - \bar{\lambda}_0}\right) = \left(I + \frac{2\sigma_0 p_2}{\lambda - \bar{\lambda}_0}\right)\left(I + \frac{2\sigma_0 Q_2}{\lambda - \bar{\lambda}_0}\right),$$

where $Q_2 = P_2 - p_2$ and $p_2 = J g_2^* g_2$; and, consequently,

$$Z(\lambda) = \left(I + \frac{2\sigma_0 p_1}{\lambda - \lambda_0}\right)\left(I + \frac{2\sigma_0 Q_1}{\lambda - \lambda_0}\right)\left(I + \frac{2\sigma_0 p_2}{\lambda - \bar{\lambda}_0}\right)\left(I + \frac{2\sigma_0 Q_2}{\lambda - \bar{\lambda}_0}\right).$$

We interchange the second and third factors. This transposition is possible, since the splitting off of elementary factors from the matrix-valued function

$$\left(I + \frac{2\sigma_0 Q_1}{\lambda - \lambda_0}\right)\left(I + \frac{2\dot{\sigma}_0 p_2}{\lambda - \bar{\lambda}_0}\right)$$

can be realized by beginning with any pole; that is, we have the equality

$$\left(I + \frac{2\sigma_0 Q_1}{\lambda - \lambda_0}\right)\left(I + \frac{2\sigma_0 p_2}{\lambda - \bar{\lambda}_0}\right) = \left(I + \frac{2\sigma_0 \hat{p}_2}{\lambda - \bar{\lambda}_0}\right)\left(I + \frac{2\sigma_0 \hat{Q}_1}{\lambda - \lambda_0}\right)$$

$$Z(\lambda) = \left(I + \frac{2\sigma_0 p_1}{\lambda - \lambda_0}\right)\left(I + \frac{2\sigma_0 \hat{p}_2}{\lambda - \bar{\lambda}_0}\right)\left(I + \frac{2\sigma_0 \hat{Q}_1}{\lambda - \lambda_0}\right)\left(I + \frac{2\sigma_0 Q_2}{\lambda - \bar{\lambda}_0}\right).$$

The factor

$$\left(I + \frac{2\sigma_0 p_1}{\lambda - \lambda_0}\right)\left(I + \frac{2\sigma_0 \hat{p}_2}{\lambda - \bar{\lambda}_0}\right)$$

is real if and only if $\hat{g}_2 = \cosh a\bar{g}_1 - \sinh a g_1$. The question reduces to whether there exists a vector $g_2 \in HP_2$ such that $\hat{g}_2 = \tilde{g}_1$, and how to choose it in $M_2 = HP_2$.

However, it is more expedient here to use another criterion for a real primary factor. From the condition

$$\left(I + \frac{2\sigma_0 p}{\lambda - \lambda_0}\right)\left(I + \frac{2\sigma_0 \tilde{p}}{\lambda - \bar{\lambda}_0}\right) = \left(I + \frac{2\sigma_0 \bar{p}}{\lambda - \bar{\lambda}_0}\right)\left(I + \frac{2\sigma_0 \bar{\tilde{p}}}{\lambda - \lambda_0}\right)$$

for the primary factor

$$z(\lambda) = \left(I + \frac{2\sigma_0 p}{\lambda - \lambda_0}\right)\left(I + \frac{2\sigma_0 \tilde{p}}{\lambda - \bar{\lambda}_0}\right)$$

to be real, it is clear that the second factor $(I + 2\sigma_0\tilde{p}/(\lambda - \bar{\lambda}_0))$ has the form $(I + 2\sigma_0\bar{p}/(\lambda - \bar{\lambda}_0))$ when it is transposed to the first place. It is not hard to see that in the case of factors of rank 1 this condition is not only necessary but also sufficient for $z(\lambda)$ to be real.

With this criterion for $z(\lambda)$ to be real in mind, we proceed as follows.

1. In $(I + 2\sigma_0 P_1/(\lambda - \lambda_0))$ we split off the primary factor $(I + 2\sigma_0 p_1/(\lambda - \lambda_0))$ generated by g_1. Then

$$Z(\lambda) = \left(I + \frac{2\sigma_0 p_1}{\lambda - \lambda_0}\right)\left(I + \frac{2\sigma_0 Q_1}{\lambda - \lambda_0}\right)\left(I + \frac{2\sigma_0 p_2}{\lambda - \bar{\lambda}_0}\right).$$

2. We transpose the last factor $(I + 2\sigma_0 P_2/(\lambda - \bar{\lambda}_0))$ into the first place:

$$Z(\lambda) = \left\{\left(I + \frac{2\sigma_0 p_1}{\lambda - \lambda_0}\right)\left(I + \frac{2\sigma_0 Q_1}{\lambda - \lambda_0}\right)\right\}\left(I + \frac{2\sigma_0 P_2}{\lambda - \bar{\lambda}_0}\right)$$

$$= \left(I + \frac{2\sigma_0 \hat{P}_2}{\lambda - \bar{\lambda}_0}\right)\left\{\left(I + \frac{2\sigma_0 \hat{p}_1}{\lambda - \lambda_0}\right)\left(I + \frac{2\sigma_0 \hat{Q}_1}{\lambda - \lambda_0}\right)\right\}.$$

Since $Z(\lambda)$ is real,

$$\left(I + \frac{2\sigma_0 \hat{P}_2}{\lambda - \bar{\lambda}_0}\right) = \left(I + \frac{2\sigma_0 \overline{P}_1}{\lambda - \bar{\lambda}_0}\right).$$

3. In this factor we single out the primary factor $(I + 2\sigma_0\bar{p}_1/(\lambda - \bar{\lambda}_0))$ generated by $\bar{g}_1$. Then

$$Z(\lambda) = \left(I + \frac{2\sigma_0 \bar{p}_1}{\lambda - \bar{\lambda}_0}\right)\left(I + \frac{2\sigma_0 \overline{Q}_1}{\lambda - \bar{\lambda}_0}\right)\left\{\left(I + \frac{2\sigma_0 \hat{p}_1}{\lambda - \lambda_0}\right)\left(I + \frac{2\sigma_0 \hat{Q}_1}{\lambda - \lambda_0}\right)\right\}.$$

4. We transpose the factor $(I + 2\sigma_0 \overline{Q}_1/(\lambda - \bar{\lambda}_0))$ to the last place:

$$Z(\lambda) = \left(I + \frac{2\sigma_0 \bar{p}_1}{\lambda - \bar{\lambda}_0}\right)\left(I + \frac{2\sigma_0 \hat{\bar{p}}_1}{\lambda - \lambda_0}\right)\left(I + \frac{2\sigma_0 \hat{Q}_1}{\lambda - \lambda_0}\right)\left(I + \frac{2\sigma_0 \hat{\bar{Q}}_1}{\lambda - \bar{\lambda}_0}\right).$$

To see that the primary factor

$$\left(I + \frac{2\sigma_0 \bar{p}_1}{\lambda - \bar{\lambda}_0}\right)\left(I + \frac{2\sigma_0 \hat{\bar{p}}_1}{\lambda - \lambda_0}\right)$$

is real, we interchange $(I + 2\sigma_0 \bar{p}_1/(\lambda - \bar{\lambda}_0))$ and $(I + 2\sigma_0 \hat{\bar{p}}_1/(\lambda - \lambda_0))$. We get

$$\left(I + \frac{2\sigma_0 \bar{p}_1}{\lambda - \bar{\lambda}_0}\right)\left(I + \frac{2\sigma_0 \hat{\bar{p}}_1}{\lambda - \lambda_0}\right) = \left(I + \frac{2\sigma_0 \hat{\bar{p}}_1}{\lambda - \lambda_0}\right)\left(I + \frac{2\sigma_0 \bar{p}_1}{\lambda - \bar{\lambda}_0}\right).$$

It is easy to see that $p_1 = \hat{\bar{p}}.(^1)$
Thus,

$$\left(I + \frac{2\sigma_0 \bar{p}_1}{\lambda - \bar{\lambda}_0}\right)\left(I + \frac{2\sigma_0 \hat{p}_1}{\lambda - \lambda_0}\right) = \left(I + \frac{2\sigma_0 p_1}{\lambda - \lambda_0}\right)\left(I + \frac{2\sigma_0 \hat{\bar{p}}_1}{\lambda - \bar{\lambda}_0}\right),$$

and, according to the criterion mentioned, this means that the split-off factor

$$\left(I + \frac{2\sigma_0 \bar{p}_1}{\lambda - \bar{\lambda}_0}\right)\left(I + \frac{2\sigma_0 \hat{p}_1}{\lambda - \lambda_0}\right)$$

is a primary real factor. We have thereby proved

THEOREM 4. *Suppose that an elementary real factor is given*

$$Z(\lambda) = \left(I + \frac{2\sigma_0 P_1}{\lambda - \lambda_0}\right)\left(I + \frac{2\sigma_0 P_2}{\lambda - \bar{\lambda}_0}\right), \qquad Z(\lambda) = \overline{Z(\bar{\lambda})}.$$

For any properly normalized basis vector g_1 in the subspace HP_1 the primary real factor

$$Z(\lambda) = \left(I + \frac{2\sigma_0 p_1}{\lambda - \lambda_0}\right)\left(I + \frac{2\sigma_0 \tilde{p}_1}{\lambda - \bar{\lambda}_0}\right)$$

generated by this basis vector splits off from $Z(\lambda)$, where $p_1 = J g_1^* g_1$, $\tilde{p}_1 = J \tilde{g}_1^* \tilde{g}$, and $\tilde{g}_1 = \cosh a \bar{g}_1 - \sinh a g_1$.

Since the remainder

$$\left(I + \frac{2\sigma_0 \hat{Q}_1}{\lambda - \lambda_0}\right)\left(I + \frac{2\sigma_0 \hat{\bar{Q}}_1}{\lambda - \bar{\lambda}_0}\right)$$

(^{1})This is a consequence of the following two properties of the transpositions:

1. If $b_{\lambda_0}(\lambda)b_{\mu_0}(\lambda) = \hat{b}_{\mu_0}(\lambda)\hat{b}_{\lambda_0}(\lambda) = \hat{\hat{b}}_{\lambda_0}(\lambda)\hat{\hat{b}}_{\mu_0}(\lambda)$, where $\lambda_0 \neq \mu_0$, then $b_{\lambda_0}(\lambda) = \hat{\hat{b}}_{\lambda_0}(\lambda)$.

2. If $b_{\lambda_0}(\lambda)\{b_{\mu_0}(\lambda)b_{\nu_0}(\lambda)\} = \{b_{\mu_0}\widehat{(\lambda)b_{\nu_0}}(\lambda)\}\hat{b}_{\lambda_0}(\lambda)$ and

$$b_{\lambda_0}(\lambda)\{b_{\mu_0}(\lambda)b_{\nu_0}(\lambda)\} = \{b_{\lambda_0}(\lambda)b_{\mu_0}(\lambda)\}b_{\nu_0}(\lambda) = \{\hat{b}_{\mu_0}(\lambda)\hat{b}_{\lambda_0}(\lambda)\}b_{\nu_0}(\lambda)$$

$$= \hat{b}_{\mu_0}(\lambda)\{\hat{b}_{\lambda_0}(\lambda)b_{\nu_0}(\lambda)\} = \hat{b}_{\mu_0}(\lambda)\{\hat{b}_{\lambda_0}(\lambda)\hat{b}_{\lambda_0}(\lambda)\} = \{\hat{b}_{\mu_0}(\lambda)\hat{b}_{\nu_0}(\lambda)\}\hat{\hat{b}}_{\lambda_0}(\lambda)$$

where $\lambda_0 \neq \mu_0$, ν_0, then $\hat{b}_{\lambda_0}(\lambda) = \hat{\hat{b}}_{\lambda_0}(\lambda)$.

where $r(\hat{\hat{Q}}_1) = r(P_1) - 1$, is an elementary real factor, the process of splitting off primary factors can be continued.

Repeating the splitting-off process $m - 1$ times $[r(P_1) = m]$, we arrive at a decomposition of an elementary real factor into a product of m primary factors.

Thus, a real rational matrix-valued function $W(\lambda)$ that is J-expansive in the right half-plane and J-unitary on the imaginary axis can be decomposed into a product of primary real factors.

BIBLIOGRAPHY

1. V. P. Potapov, *The multiplicative structure of J-nonexpansive matrix-valued functions*, Trudy Moskov. Mat. Obshch. **4** (1955), 125–236; English transl. in Amer. Math. Soc. Transl. (2) **15** (1960).

Translated by H. H. McFADEN

Amer. Math. Soc. Transl.
(2) Vol. **138**, 1988

General Theorems on the Structure
and the Splitting Off of Elementary Factors
of Analytic Matrix-Valued Functions

UDC 519.210

V. P. POTAPOV

1°. In this paper we study the question of the multiplicative structure of single-valued analytic matrix-valued functions on the right half-plane. It is assumed that the only singular points of the matrix-valued function $W(\lambda)$ are poles and that it has the following three properties:

1. $W^*(\lambda)JW(\lambda) - J \geq 0$ (Re $\lambda > 0$),
2. $\overline{W(\bar{\lambda})} \equiv W(\lambda)$,
3. $W'(\lambda)jJW(\lambda) \equiv jJ$,

where

$$J = \begin{pmatrix} 0 & I_n \\ I_n & 0 \end{pmatrix}, \qquad j = \begin{pmatrix} I_n & 0 \\ 0 & -I_n \end{pmatrix}.$$

Such a matrix-valued function is called a *circuit function*. If, in addition, the limit values $W(i\tau)$ satisfy the condition

4. $W^*(i\tau)JW(i\tau) \equiv J,$

then $W(\lambda)$ is said to be *circuit reactive*.

Recall that, according to [1]–[3], a matrix W satisfying condition (1) is said to be *J-noncontractive*, and a matrix W satisfying (4) is said to be *J-unitary*.

Here we give an exhaustive description of the simplest circuit reactive matrix-valued functions $b(\lambda)$—the so-called primary factors—and we prove that an arbitrary rational circuit reactive matrix-valued function can be represented as a product of finitely many primary factors. This result, which has fundamental significance for the theory of electrical circuit design, was first established by Kovalishina [3]. However, the description of primary factors given there is not transparent enough.

1980 *Mathematics Subject Classification* (1985 *Revision*). Primary 15A23, 30G30.
Translation of Akad. Nauk Armyan. SSR Dokl. **48** (1969), 257–263; MR **57** #16627.

The investigation of the multiplicative structure of a circuit matrix-valued function is based on some general theorems which are of independent interest.

2°. THEOREM 1. *A J-noncontractive matrix-valued function $W(\lambda)$ which is regular at the points $\lambda_1,\ldots,\lambda_m$, $\operatorname{Re}\lambda_l \neq 0$, $l = 1,\ldots,m$, satisfies the following inequality:*

$$\left\| \frac{W^*(\lambda_k)JW(\lambda_l) - J}{\bar{\lambda}_k + \lambda_l} \right\|_{k,l=1}^{m} \geq 0. \tag{1}$$

In the case when λ_l lies in the left half-plane we understand $W(\lambda_l)$ to be the extension "by symmetry", namely,

$$W(\lambda_l) = JW^{*-1}(-\bar{\lambda}_l)J,$$

independently of whether or not $W(\lambda)$ extends to the left half-plane. The ambiguity at a pair of points symmetric with respect to the imaginary axis is revealed according to the usual rules.

Inequality (1) is a generalization of the Schwarz lemma. The corresponding inequality for matrix-valued functions with nonnegative imaginary part follows easily from the well-known integral representation.

A reasonable passage to the limit in (1) leads to the following theorem, which plays a basic role in the decomposition of $W(\lambda)$ into factors.

THEOREM 2. *A J-noncontractive matrix-valued function*

$$W(\lambda) = \frac{2\sigma_l C_l}{(\lambda - \lambda_l)^{S_l}} + \frac{b_l}{(\lambda - \lambda_l)^{S_{l-1}}} + \cdots$$

with poles at the points $\lambda_1,\ldots,\lambda_m$ satisfies the inequality

$$\begin{pmatrix} A & B \\ B^* & c \end{pmatrix} = \left[\begin{array}{ccc:c} \dfrac{4\sigma_1\sigma_1}{\bar{\lambda}_1 + \lambda_1}C_1^* JC_1 & \cdots & \dfrac{4\sigma_1\sigma_m}{\bar{\lambda}_1 + \lambda_m}C_1^* JC_m & \dfrac{2\sigma_1 C_1^*}{\bar{\lambda}_1 - \bar{\mu}} \\[2mm] \vdots & & \vdots & \vdots \\[2mm] \dfrac{4\sigma_m\sigma_1}{\bar{\lambda}_1 + \lambda_1}C_m^* JC_1 & \cdots & \dfrac{4\sigma_m\sigma_m}{\bar{\lambda}_m + \lambda_m}C_m^* JC_m & \dfrac{2\sigma_m C_m^*}{\bar{\lambda}_m - \bar{\mu}} \\[1mm] \hdashline \dfrac{2\sigma_1 C_1}{\lambda_1 - \mu} & \cdots & \dfrac{2\sigma_m C_m}{\lambda_m - \mu} & \dfrac{W(\mu)JW^*(\mu) - J}{\mu + \bar{\mu}} \end{array} \right] \geq 0. \tag{2}$$

The presence of pairs of points symmetric with respect to the imaginary axis leads to a change in the corresponding elements of the block A. Inequality (2) can be written also for the case of a pole on the imaginary axis if the leading coefficient C satisfies the condition $C^* JC = 0$. The accompanying changes in the block A, which disturb the uniformity of the expression, are not of fundamental significance and will not be discussed in what follows.

The investigation of (2) is based on the following elementary proposition.

LEMMA ON BLOCK MATRICES. *The matrix*

$$\begin{pmatrix} A & B \\ B^* & c \end{pmatrix}$$

is nonnegative-Hermitian if and only if the following three conditions hold.
1. $A \geq 0$;
2. *the equation $Ax = B$ has solutions; and*
3. *for any solution x of the latter equation*

$$c - x^* A x \geq 0. \tag{3}$$

From an arbitrary J-noncontractive matrix-valued function $W(\lambda)$ we now split off factors $b(\lambda)$ of the same nature which are as simple as possible from an analytic point of view. Their simplicity is characterized by the fact that they have *simple* poles at the given points $\lambda_1, \dots, \lambda_m$ and are J-unitary on the imaginary axis. Such factors will be called *elementary* factors. If they are normalized by the condition $b(\infty) = J$, then

$$b(\lambda) = I + \frac{2\sigma_1}{\lambda - \lambda_1} a_1 + \frac{2\sigma_2}{\lambda - \lambda_2} a_2 + \cdots + \frac{2\sigma_m}{\lambda - \lambda_m} a_m.$$

(Note: Only binomial factors $b(\lambda) = I + 2\sigma_0 a/(\lambda - \lambda_0)$ were considered in [1], [2], and [3]. The necessity of extending further is dictated by the fact that in the general case a circuit reactive elementary factor is a sum of five terms.)

In the particular case when all the residues $a_1, \dots, a_m$ are matrices of rank 1 the factor $b(\lambda)$ will be called a *primary* factor and will be written in the form

$$b(\lambda) = J + \frac{2\sigma_1}{\lambda - \lambda_1} J f_1^* g_1 + \frac{2\sigma_2}{\lambda - \lambda_2} J f_2^* g_2 + \cdots + \frac{2\sigma_m}{\lambda - \lambda_m} J f_m^* g_m,$$

where f and g are row vectors.

An elementary factor $b(\lambda)$ plays the role of the linear fractional function in the classical Schwarz lemma. We have

THEOREM 3. *A J-noncontractive matrix-valued function $W(\lambda)$ is an elementary factor if and only if in the block matrix (2) inequality (3) becomes the equality $c - x^* A x = 0$.*

The structure of an elementary factor is described by

THEOREM 4. *A matrix-valued function*

$$b(\lambda) = I + \frac{2\sigma_1}{\lambda - \lambda_1} a_1 + \frac{2\sigma_2}{\lambda - \lambda_2} a_2 + \cdots + \frac{2\sigma_m}{\lambda - \lambda_m} a_m$$

is an elementary factor if and only if the following two conditions hold:

$$\begin{bmatrix} \dfrac{4\sigma_1 \sigma_1}{\bar{\lambda}_1 + \lambda_1} a_1^* J a_1 & \cdots & \dfrac{4\sigma_1 \sigma_m}{\bar{\lambda}_1 + \lambda_m} a_1^* J a_m \\ \vdots & & \vdots \\ \dfrac{4\sigma_m \sigma_1}{\bar{\lambda}_m + \lambda_1} a_m^* J a_1 & \cdots & \dfrac{4\sigma_m \sigma_m}{\bar{\lambda}_m + \lambda_m} a_m^* J a_m \end{bmatrix} \geq 0, \tag{A}$$

$$\frac{4\sigma_k \sigma_1}{\bar{\lambda}_k + \lambda_1} a_k^* J a_1 + \frac{4\sigma_k \sigma_2}{\bar{\lambda}_k + \lambda_2} a_k^* J a_2 + \cdots + \frac{4\sigma_k \sigma_m}{\bar{\lambda}_k + \lambda_m} a_k^* J a_m = 2\sigma_k a_k^* J \tag{B}$$

$$(k = 1, 2, \dots, m).$$

A primary factor has an especially simple characterization.

THEOREM 5. *A matrix-valued function*

$$b(\lambda) = I + \frac{2\sigma_1}{\lambda - \lambda_1} J f_1^* g_1 + \frac{2\sigma_2}{\lambda - \lambda_2} J f_2^* g_2 + \cdots + \frac{2\sigma_m}{\lambda - \lambda_m} J f_m^* g_m$$

is a primary factor if and only if the vectors $f_1, \ldots, f_m$ *satisfy the "sharp" inequality*

$$\begin{bmatrix} \dfrac{4\sigma_1\sigma_1}{\bar{\lambda}_1 + \lambda_1} f_1 J f_1^* & \cdots & \dfrac{4\sigma_1\sigma_m}{\bar{\lambda}_1 + \lambda_m} f_1 J f_m^* \\ \vdots & & \vdots \\ \dfrac{4\sigma_m\sigma_1}{\bar{\lambda}_m + \lambda_1} f_m J f_1^* & \cdots & \dfrac{4\sigma_m\sigma_m}{\bar{\lambda}_m + \lambda_m} f_m J f_m^* \end{bmatrix} > 0, \qquad (\mathrm{A}')$$

and the vectors $g_1, \ldots, g_m$ *satisfy the system of equations*

$$\frac{4\sigma_k\sigma_1}{\bar{\lambda}_k + \lambda_1}(f_k J f_1^*)g_1 + \frac{4\sigma_k\sigma_2}{\bar{\lambda}_k + \lambda_2}(f_k J f_2^*)g_2$$
$$+ \cdots + \frac{4\sigma_k\sigma_m}{\bar{\lambda}_k + \lambda_m}(f_k J f_m^*)g_m = 2\sigma_k f_k \qquad (k = 1, 2, \ldots, m). \qquad (\mathrm{B}')$$

The construction of an elementary factor splitting off from a given J-noncontractive matrix-valued function $W(\lambda)$ is given in

THEOREM 6. *Suppose that the* J-*noncontractive matrix-valued function* $W(\lambda)$ *has poles at the points* $\lambda_1, \ldots, \lambda_m$,

$$W(\lambda) = \frac{2\sigma_l C_l}{(\lambda - \lambda_l)^{S_l}} + \frac{b_l}{(\lambda - \lambda_l)^{S_l - 1}} + \cdots \qquad (l = 1, 2, \ldots, m),$$

and suppose that the matrices $X_1, \ldots, X_m$ *are found from the system of equations*

$$\frac{4\sigma_k\sigma_1}{\bar{\lambda}_k + \lambda_1} C_k^* J C_1 X_1 + \frac{4\sigma_k\sigma_2}{\bar{\lambda}_k + \lambda_2} C_k^* J C_2 X_2$$
$$+ \cdots + \frac{4\sigma_k\sigma_m}{\bar{\lambda}_k + \lambda_m} C_k^* J C_m X_m = 2\sigma_k C_k^* \qquad (k = 1, 2, \ldots, m).$$

Then the following assertions are true:

1) *The matrix-valued function*

$$b(\lambda) = I + \frac{2\sigma_1}{\lambda - \lambda_1} C_1 X_1 J + \frac{2\sigma_2}{\lambda - \lambda_2} C_2 X_2 J + \cdots + \frac{2\sigma_m}{\lambda - \lambda_m} C_m X_m J$$

is an elementary factor.

2) *This factor splits off from* $W(\lambda)$ *on the left; namely,* $W(\lambda) = b(\lambda)W_1(\lambda)$.

3) *The orders of the poles* $\lambda_1, \ldots, \lambda_m$ *are lowered by 1 by this splitting off.*

The corresponding theorem for primary factors asserts that the ranks of the leading coefficients are lowered by 1.

3°. In conclusion we give a complete description of the structure of a primary reactive factor.

Since a circuit reactive matrix-valued function with a pole at a point $\lambda_0 = \sigma_0 + i\tau_0$ $(\sigma_0, \tau_0 > 0)$ must have poles at the symmetric points $-\bar{\lambda}_0$, $\bar{\lambda}_0$, and $-\lambda_0$, it is natural here to understand an elementary factor to be a pentanomial

$$b(\lambda) = I + \frac{2\sigma_0}{\lambda - \lambda_0} a_1 - \frac{2\sigma_0}{\lambda + \bar{\lambda}_0} a_2 + \frac{2\sigma_0}{\lambda - \bar{\lambda}_0} a_3 - \frac{2\sigma_0}{\lambda + \lambda_0} a_4.$$

It is not hard to see that

$$a_1 = a, \qquad a_2 = j\bar{a}j, \qquad a_3 = \bar{a}, \qquad a_4 = jaj.$$

For a primary factor the matrix a is of rank 1, $a = Jf^*g$, and the problem reduces to establishing necessary and sufficient conditions on the vector f and g for the factor $b(\lambda)$ to be a circuit reactive matrix-valued function.

If in the formulation of Theorem 5 we introduce the corresponding corrections ensuring that there are two pairs of poles symmetric with respect to the imaginary axis, we get

THEOREM 7. *Suppose that the $2n$-dimensional vector f is such that the numbers*

$$\alpha = fJf^*, \qquad \gamma = \frac{\sigma_0}{\lambda_0}fJ\bar{f}^*, \qquad \delta = i\frac{\sigma_0}{\tau_0}fjJf^*$$

are connected by the relations

$$|\gamma| < \alpha, \qquad |\delta| < \alpha. \tag{4}$$

Then the matrix inequality

$$\begin{pmatrix} \alpha & \theta & \gamma & \delta \\ \bar{\theta} & \alpha & \delta & \bar{\gamma} \\ \bar{\gamma} & \delta & \alpha & \bar{\theta} \\ \delta & \gamma & \theta & \alpha \end{pmatrix} > 0 \tag{5}$$

with respect to the complex parameter θ has infinitely many solutions; and for each such θ the vector g determined from the system of equations

$$\begin{aligned}
\alpha g + \theta \bar{g}j + \gamma \bar{g} + \delta gj &= f, \\
\bar{\theta} g + \alpha \bar{g}j + \delta \bar{g} + \bar{\gamma} gj &= \bar{f}j, \\
\bar{\gamma} g + \delta \bar{g}j + \alpha \bar{g} + \bar{\theta} gj &= \bar{f}, \\
\delta g + \gamma \bar{g}j + \theta \bar{g} + \alpha gj &= fj,
\end{aligned} \tag{6}$$

together with the vector f, gives rise to a primary reactive factor

$$b(\lambda) = I + \frac{2\sigma_0}{\lambda - \lambda_0}Jf^*g - \frac{2\sigma_0}{\lambda + \bar{\lambda}_0}jJ\bar{f}^*\bar{g}j + \frac{2\sigma_0}{\lambda - \bar{\lambda}_0}Jf^*\bar{g} - \frac{2\sigma_0}{\lambda + \lambda_0}jJf^*gj. \tag{7}$$

Conversely, if $b(\lambda)$ is a primary reactive factor, then the vectors f and g are connected by the equations (6), in which the numbers α, θ, γ, and δ satisfy the "sharp" matrix inequality (5), and

$$\alpha = fJf^*, \qquad \gamma = \frac{\sigma_0}{\lambda_0}fJ\bar{f}^*, \qquad \delta = i\frac{\sigma_0}{\tau_0}fjJf^*.$$

We do not dwell here on the cases when the poles of a primary reactive factor lie on the real or the imaginary axis, since its structure is much simpler.

Bibliography

1. V. P. Potapov, *The multiplicative structure of J-nonexpansive matrix-valued functions*, Trudy Moskov. Mat. Obshch. **4** (1955), 125–236; English transl. in Amer. Math. Soc. Transl. (2) **15** (1960).

2. I. V. Kovalishina and V. P. Potapov, *The multiplicative structure of analytic real J-expansive matrix-valued functions*, Izv. Akad. Nauk Armyan. SSR Ser. Fiz.-Mat. Nauk **18** (1965), no. 6, 3–10; English transl. in this volume.

3. I. V. Kovalishina, *Multiplicative structure of analytic reactive matrix-valued functions*, Izv. Akad. Nauk Armyan. SSR Ser. Mat. **1** (1966), 138–146. (Russian)

Translated by H. H. McFADEN

Amer. Math. Soc. Transl.
(2) Vol. **138**, 1988

An Indefinite Metric in the Nevanlinna-Pick Problem
UDC 519.210

I. V. KOVALISHINA AND V. P. POTAPOV

In this note we consider the Nevanlinna-Pick problem [1] in a matrix aspect. Namely, given a sequence of points $\lambda_1, \lambda_2, \ldots$ $(\lambda_j + \bar{\lambda}_j > 0)$ in the open right half-plane and a sequence of $m \times m$ matrices $w_1, w_2, \ldots$ $(w_j + w_j^* > 0)$ in the open right matrix half-plane, find a positive matrix-valued function $w = w(\lambda)$ that is holomorphic in the right half-plane, satisfies the condition

$$\operatorname{Re} w(\lambda) = \tfrac{1}{2}[w(\lambda) + w^*(\lambda)] \geq 0$$

and is such that $w(\lambda_1) = w_1$, $w(\lambda_2) = w_2, \ldots$.

Here we establish a very close connection between this problem and the theory of J-expansive matrix-valued functions [2].

The investigation is based on the Schwarz-Pick inequality for positive matrix-valued functions:

$$\left\| \frac{w(\lambda_j) + w^*(\lambda_k)}{\lambda_j + \bar{\lambda}_k} \right\|_{j,k=1}^{n} \geq 0. \tag{1}$$

As usual, we first solve the truncated problem. We have the

THEOREM. *A matrix-valued function $w(\lambda)$ is a solution of the Nevanlinna-Pick problem for a finite number of pairs λ_j, w_j $(j = 1, \ldots, n)$ if and only if it satisfies the basic matrix inequality*

$$\left| \begin{array}{ccc|c} \dfrac{w_1 + w_1^*}{\lambda_1 + \bar{\lambda}_1} & \cdots & \dfrac{w_1 + w_n^*}{\lambda_1 + \bar{\lambda}_n} & \dfrac{w_1 + w^*(\lambda)}{\lambda_1 + \bar{\lambda}} \\ \cdots & \cdots & \cdots & \cdots \\ \dfrac{w_n + w_1^*}{\lambda_n + \bar{\lambda}_1} & \cdots & \dfrac{w_n + w_n^*}{\lambda_n + \bar{\lambda}_n} & \dfrac{w_n + w^*(\lambda)}{\lambda_n + \bar{\lambda}} \\ \hline \dfrac{w(\lambda) + w_1^*}{\lambda + \bar{\lambda}_1} & \cdots & \dfrac{w(\lambda) + w_n^*}{\lambda + \bar{\lambda}_n} & \dfrac{w(\lambda) + w^*(\lambda)}{\lambda + \bar{\lambda}} \end{array} \right| \geq 0. \tag{2}$$

1980 *Mathematics Subject Classification* (1985 *Revision*). Primary 30E05, 30G30.
Translation of Akad. Nauk Armyan. SSR Dokl. **59** (1974), 17–22; MR **53** #13577.

Under the assumption that the block

$$
A = \begin{vmatrix} \dfrac{w_1 + w_1^*}{\lambda_1 + \bar{\lambda}_1} & \cdots & \dfrac{w_1 + w_n^*}{\lambda_1 + \bar{\lambda}_n} \\ \cdots & \cdots & \cdots \\ \dfrac{w_n + w_1^*}{\lambda_n + \bar{\lambda}_1} & \cdots & \dfrac{w_n + w_n^*}{\lambda_n + \bar{\lambda}_n} \end{vmatrix}
$$

is nonsingular,[1] the solution of (2) can be written as a linear fractional transformation of an arbitrary positive matrix-valued function $\omega(\lambda)$,

$$
w(\lambda) = [\omega(\lambda)b(\lambda) + d(\lambda)]^{-1}[\omega(\lambda)a(\lambda) + c(\lambda)],
$$

whose coefficient matrix is a group elementary factor [3]; that is, a matrix-valued function J-expansive in the right half-plane and J-unitary on the imaginary axis:

$$
B_n(\lambda) = I + \sum_{j=1}^{n} \frac{2\sigma_j a_j}{\lambda - \lambda_j} = \begin{bmatrix} a(\lambda) & b(\lambda) \\ c(\lambda) & d(\lambda) \end{bmatrix}, \tag{3}
$$

where

$$
J = \begin{pmatrix} 0 & I_m \\ I_m & 0 \end{pmatrix},
$$

$$
2\sigma_j a_j = \sum_{k=1}^{n} \begin{bmatrix} I \\ w_k^* \end{bmatrix} h_{kj}[I, w_j]J,
$$

$$
\|h_{jk}\|_{j,k=1}^{n} = A^{-1}, \qquad 2\sigma_j = \lambda_j + \bar{\lambda}_j.
$$

We consider the stepwise solution of the problem in parallel. The following facts are established:

1. *Each J-nonnegative projection $P = P^2$, $PJ \geq 0$, of full rank[2] can be written in the form*

$$
P = \begin{bmatrix} (w_0 + w_0^*)^{-1}w_0 & (w_0 + w_0^*)^{-1} \\ w_0^*(w_0 + w_0^*)^{-1}w_0 & w_0^*(w_0 + w_0^*)^{-1} \end{bmatrix}, \qquad w_0 + w_0^* > 0.
$$

2. *If*

$$
L_0(\lambda) = I + \frac{2\sigma_0}{\lambda - \lambda_0}P = \begin{bmatrix} a(\lambda) & b(\lambda) \\ c(\lambda) & d(\lambda) \end{bmatrix}
$$

is an elementary factor of full rank, and $\omega(\lambda)$ is an arbitrary positive matrix-valued function, then the linear fractional transformation

$$
w(\lambda) = [\omega(\lambda)b(\lambda) + d(\lambda)]^{-1}[\omega(\lambda)a(\lambda) + c(\lambda)] = L_0(\lambda)\{\omega(\lambda)\}
$$

determines a positive matrix-valued function $w(\lambda)$ that is the general solution of the interpolation problem $w(\lambda_0) = w_0$ with a single node.

3. *The general form of positive matrix-valued functions satisfying any finite number of first conditions $w(\lambda_1) = w_1, \ldots, w(\lambda_n) = w_n$ can be represented as a*

---

[1] In the scalar case the singularity of the block A leads to a unique solution which is a rational positive function. The question of the singularity of A in the matrix case will not be discussed here.

[2] That is, of maximal rank, which equals m in the given case.

superposition of linear fractional transformations

$$w(\lambda) = L_1(\lambda)\{L_2(\lambda)\{\ldots\{L_n(\lambda)\{\omega_n(\lambda)\}\}\ldots\}\},$$

where $\omega_n(\lambda)$ is an arbitrary positive matrix-valued function.

The coefficient matrix of the resulting linear fractional transformation is equal to the product of elementary factors of full rank

$$L_n(\lambda)L_{n-1}(\lambda)\cdots L_2(\lambda)L_1(\lambda),$$

$$L_i(\lambda) = I + \frac{2\sigma_i}{\lambda - \lambda_i}P_i,$$

$$P_i = \begin{bmatrix} (w_i^{(i)} + w_i^{(i)*})^{-1}w_i^{(i)} & (w_i^{(i)} + w_i^{(i)*})^{-1} \\ w_i^{(i)*}(w_i^{(i)} + w_i^{(i)*})^{-1}w_i^{(i)} & w_i^{(i)*}(w_i^{(i)} + w_i^{(i)*})^{-1} \end{bmatrix}, \qquad (4)$$

where the matrices $w_i^{(i)}$—the so-called Schur parameters—are found successively by the formulas

$$w_i^{(i)} = L_{i-1}^{-1}(\lambda_i)\{L_{i-2}^{-1}(\lambda_i)\{\ldots\{L_1^{-1}(\lambda_i)\{w_i\}\}\ldots\}\}, \qquad (i = 2, 3, \ldots, n),$$

$$(w_1^{(1)} = w_1).$$

Thus, to the statement of any Nevanlinna-Pick problem there corresponds an infinite product of binomial factors of full rank

$$\cdots L_n(\lambda)L_{n-1}(\lambda)\cdots L_2(\lambda)L_1(\lambda),$$

whose nth partial product(^{3}) $B_n(\lambda) = L_n(\lambda)\cdots L_1(\lambda)$ is the coefficient matrix of that linear fractional transformation of an arbitrary positive function which gives the general solution of the truncated problem. The converse assertion is also true. Indeed, take an arbitrary infinite product of binomial factors of full rank:

$$\cdots L_n(\lambda)\cdots L_2(\lambda)L_1(\lambda), \qquad L_i(\lambda) = I + \frac{2\sigma_i}{\lambda - \lambda_i}P_i,$$

where the λ_i are distinct. The projection P_i of each such factor determines a positive matrix $w_i^{(i)}$. From $w_i^{(i)}$ we construct the matrices

$$w_i = L_1(\lambda_i)\{L_2(\lambda_i)\{\ldots\{L_{i-1}(\lambda_i)\{w_i^{(i)}\}\}\cdots\}\}$$

and we formulate the Nevanlinna-Pick problem: find a positive matrix-valued function $w(\lambda)$ satisfying the conditions $w(\lambda_j) = w_j$ $(j = 1, 2, \ldots)$.

It is thereby shown that the Nevanlinna-Pick problem is adequate to the study of an infinite product of binomial factors of full rank.

The subsequent investigation is connected with a consideration of Weyl disks. In terms of $B_n(\lambda)$ inequality (2) can be rewritten in the form

$$[w(\lambda), I]\frac{B_n^{-1}(\lambda)JB_n^{*-1}(\lambda)}{\lambda + \bar{\lambda}}\begin{bmatrix} w^*(\lambda) \\ I \end{bmatrix} \geq 0$$

(^{3})The notation $L_n(\lambda)\cdots L_1(\lambda) = B_n(\lambda)$ is not accidental. It is proved that the matrix-valued function $B_n(\lambda)$ in (3) decomposes into a product of factors $L_i(\lambda)$ of the form (4).

or

$$[w(\lambda), I] \begin{pmatrix} -R_n & S_n^* \\ S_n & -T_n \end{pmatrix} \begin{bmatrix} w^*(\lambda) \\ I \end{bmatrix} \geq 0, \qquad (5)$$

where

$$B_n^{-1}(\lambda) J B_n^{*-1}(\lambda) = \begin{pmatrix} -R_n & S_n^* \\ S_n & -T_n \end{pmatrix} = W_n$$

is a so-called *Weyl matrix*. Inequality (5) means that for fixed λ the set of solutions $w(\lambda)$ of (2) fills the matrix Weyl disk of $B_n(\lambda)$:

$$w(\lambda) = S_n R_n^{-1} + \sqrt{S_n R_n^{-1} S_n^* - T_n} \cdot u \cdot \frac{1}{\sqrt{R_n}}, \qquad uu^* \leq 1.$$

A study of the behavior of the Weyl disks as n grows leads to the following results: the centers $S_n R_n^{-1}$ tend to a finite limit, the radii $\rho_g^{(n)} = S_n R_n^{-1} S_n^* - T_n$ and $\rho_d^{(n)} = R_n^{-1}$ are monotonically decreasing, and each Weyl disk is imbedded in the preceding one.

By the main theorem of Orlov in [5], the ranks of the left-hand and right-hand radii of the limit disk do not depend on the choice of $\lambda \neq \lambda_j$.

We say that the Nevanlinna-Pick problem is *completely indeterminate* if both radii have nonsingular limits.

Here is a theorem that generalizes the known criterion of Denjoy (this becomes more transparent if it is considered that the independent variable ς varies in the unit disk):

The Nevanlinna-Pick problem is completely indeterminate if and only if the series $\sum_1^\infty (1 - |\varsigma_j|) P_j J \; (|\varsigma_j| < 1)$ converges.

This condition is necessary and sufficient for the convergence of a Blaschke-Potapov product of binomial factors normalized at the point $\varsigma = 0$ to the J-modulus [2].

And, finally, a theorem has been proved on the structure of the radii of the Weyl disk for any truncated Nevanlinna-Pick problem. By using this theorem it was possible to prove the following assertions for the limit radii:

1. *If the left-hand radius is nonsingular, then so is the right-hand radius.*

2. *If the right-hand radius is nonsingular, then the left-hand radius is either nonsingular or equal to zero.*

3. *It is possible to construct a Nevanlinna-Pick problem with singular left-hand and right-hand radii of any specified ranks.*

Organic connections between the theory of analytic matrix-valued functions and other classical problems will be dealt with in separate investigations.

BIBLIOGRAPHY

1. Rolf Nevanlinna, *Über beschränkte analytische Funktionen*, Ann. Acad. Sci. Fenn. Ser. A **32** (1929), no. 7.

2. V. P. Potapov, *The multiplicative structure of J-nonexpansive matrix-valued functions*, Trudy Moskov. Mat. Obshch. **4** (1955), 125–236; English transl. in Amer. Math. Soc. Transl. (2) **15** (1960).

3. _____, *General theorems on the structure and the splitting off of elementary factors of analytic matrix-valued functions*, Akad. Nauk Armyan. SSR Dokl. **48** (1969), 257–263; English transl. in this volume.

4. A. V. Efimov and V. P. Potapov, *J-expansive matrix-valued functions and their role in the analytic theory of electric circuits*, Uspekhi Mat. Nauk **28** (1973), no. 1 (169), 65–130; English transl. in Russian Math. Surveys **28** (1973).

5. S. A. Orlov, *A theorem on the stabilization of nested matrix disks that depend analytically on a parameter*, All-Union Conf. Theory of Functions of a Complex Variable, Abstracts of Reports, Fiz.-Tekhn. Inst. Nizkikh Temperatur Akad. Nauk Ukrain. SSR, Kharkov, 1971, pp. 163–167. (Russian)*

Translated by H. H. McFADEN

Editor's note.* These results were published in S. A. Orlov, *Nested matrix disks analytically depending on a parameter, and theorems on the invariance of ranks of radii of limiting disks*, Izv. Akad. Nauk SSSR Ser. Mat. **10 (1976), 593–644; English transl. in Math. USSR-Izv. **10** (1976).

Amer. Math. Soc. Transl.
(2) Vol. **138**, 1988

Linear Fractional Transformations of Matrices

UDC 517.9

V. P. POTAPOV

This paper bears a methodological character. In it we present theorems on linear fractional transformations of square matrices having certain special properties in a form as close as possible to the needs of J-theory (see [1]–[7]).

Linear fractional transformations of a matrix disk were first considered by Siegel [8]. A general theory was then constructed by Hua Luo-geng [9] in connection with automorphisms of certain groups. However, the role of an indefinite metric in the study of transformations of certain matrix classes important for applications was apparently first noted in [1].

Introduction

Let us consider a linear fractional transformation

$$y = (ax + b)(cx + d)^{-1}, \tag{1}$$

where a, b, c, and d are constant $n \times n$ matrices, and x and y are variable $n \times n$ matrices. We associate the $2n \times 2n$ matrix of coefficients

$$A = \begin{pmatrix} a & b \\ c & d \end{pmatrix}.$$

This transformation makes sense for those and only those matrices x such that the matrix $cx + d$ is nonsingular. The set of such matrices is denoted by $\mathcal{R}_x$, and the set of corresponding matrices y by $\mathcal{R}_y$.

Note first of all that $\mathcal{R}_x$ is nonempty if and only if the rank r of the $n \times 2n$ rectangular matrix $[c, d]$ is equal to n. Indeed, if $r < n$, then the rows of the matrix $[c, d]$ are linearly dependent, and then there exists a row vector $f \neq 0$ such that $fc = 0$ and $fd = 0$. Therefore $f(cx + d) = 0$ for any matrix x, i.e., the matrix $cx + d$ is singular, and $\mathcal{R}_x$ is empty.

1980 *Mathematics Subject Classification* (1985 *Revision*). Primary 15A99.

Translation of Studies in the Theory of Operators and Their Applications (V. A. Marchenko and V. Ya. Golodets, editors), "Naukova Dumka", Kiev, 1979, pp. 75–97; MR **81f**: 15023.

Conversely, suppose that $\mathcal{R}_x$ is empty, i.e., $cx + d$ is singular for any matrix x. Let $x = c^*y^{-1}$, where the nonsingular matrix y is chosen so that $dy = dd^*$. By our assumption, there exists a vector $f \neq 0$ such that

$$0 = f(cx + d) = f(cc^* + dd^*)y^{-1},$$

and we get successively that

$$f(cc^* + dd^*) = 0, \qquad f(cc^* + dd^*)f^* = 0,$$
$$fcc^*f^* = 0, \qquad fdd^*f^* = 0, \qquad fc = 0, \qquad fd = 0, \qquad f[c, d] = 0.$$

Thus, the rows of $[c, d]$ are linearly dependent, i.e., its rank r is less than n.

It is natural to require in what follows that the transformation (1) establish a one-to-one correspondence between the sets $\mathcal{R}_x$ and $\mathcal{R}_y$. We prove that this is so if and only if the coefficient matrix $A = \left(\begin{smallmatrix} a & b \\ c & d \end{smallmatrix}\right)$ is nonsingular. Indeed, assume that A is singular. Then there exists a nonzero column vector

$$\begin{pmatrix} f^* \\ g^* \end{pmatrix}$$

which annihilates the matrix A from the right, i.e., $af^* + bg^* = 0$ and $cf^* + dg^* = 0$.

If now x_1 is a matrix such that (1) makes sense and h is a sufficiently small vector, then the transformation makes sense also for the matrix

$$x_2 = (x_1 + f^*h)(I + g^*h)^{-1}$$

and we get that

$$\begin{aligned}
(ax_2 + b)(cx_2 + d)^{-1} &= [a(x_1 + f^*h)(I + g^*h)^{-1} + b] \\
&= [ax_1 + b + af^*h + bg^*h] = (ax_1 + b)(cx_1 + d)^{-1},
\end{aligned}$$

which attests that the one-to-oneness fails.

Conversely, suppose that A is nonsingular. From $y = (ax + b)(cx + d)^{-1}$ we get successively that

$$y(cx + d) = ax + b, \qquad yd - b = (-yc + a)x.$$

Let us show that $a - yc$ is nonsingular. Indeed, otherwise there exists a vector $f \neq 0$ such that $f(a - yc) = 0$. But then $f(b - yd) = 0$. Setting $fy = g$, we get that the $2n$-dimensional vector (f, g) annihilates A, which is impossible. Accordingly, $a - cy$ is nonsingular, and

$$x = (a - yc)^{-1}(yd - b). \tag{1'}$$

Thus, a given matrix y in $\mathcal{R}_y$ uniquely determines a matrix x in $\mathcal{R}_x$.

Starting from the requirement that the correspondence between $\mathcal{R}_x$ and $\mathcal{R}_y$ be one-to-one, we consider linear fractional transformations (1) with nonsingular A everywhere in what follows.

The last formula indicates that along with transformations of the form (1) we should also consider transformations of the form

$$y = (xc_1 + d_1)^{-1}(xa_1 + b_1). \tag{2}$$

It is convenient to associate with such transformations the matrix

$$B = \begin{pmatrix} a_1 & c_1 \\ b_1 & d_1 \end{pmatrix}.$$

The rectangular matrix $[c, d]$ is said to be *nondegenerate* if it has the maximal possible rank n. Obviously, the pair $[c, d]$ is nondegenerate if and only if the nonnegative matrix $cc^* + dd^*$ is nonsingular.

This implies, in particular, that the pair $[c, d]$ is nondegenerate simultaneously with the pair $[TcU, TdV]$, where T is an arbitrary nonsingular matrix, and U and V are arbitrary unitary matrices.

The following Theorem 1 has a less trivial character. This theorem has the role of enabling us to prove various kinds of identities by working with elementary (real symmetric!) matrices of the form

$$x = \begin{pmatrix} \rho_1 & & \\ & \ddots & \\ & & \rho_n \end{pmatrix}, \qquad x = \begin{pmatrix} 0 & 1 & 0 & & & \\ 1 & 0 & 1 & & & \\ 0 & 1 & 0 & \ddots & & \\ & & \ddots & \ddots & 1 \\ & & & 1 & 0 \end{pmatrix}.$$

THEOREM 1. *If $[c, d]$ is a nondegenerate pair, then the collection of real symmetric matrices x such that $cx + d$ is nonsingular is dense in the set of all real symmetric matrices.*

PROOF. The theorem asserts that for any given real symmetric matrix x_0 there exist real symmetric matrices x arbitrarily close to it such that $\det(cx+d) \neq 0$. Assume not, i.e., there exist a real symmetric matrix x_0 and a neighborhood of it such that $\det(cx+d) \equiv 0$ for all real symmetric matrices x in this neighborhood. Then (by the principle of analytic continuation) the matrix $cx + d$ is singular for all symmetric matrices, also complex ones. The same property must be enjoyed by the matrix

$$CZ + D = Tc\overline{U}U'xU + T\,dU \qquad (C = Tc\overline{U}, D = TdU, Z = U'xU)$$

for any symmetric matrix Z. Here T is nonsingular and U is unitary. The arbitrariness in the choice of T and U is used to give D as simple a form as possible. Let T and U be such that

$$TdU = \begin{pmatrix} 0 & 0 \\ 0 & I \end{pmatrix} = [0, \ldots, 0, J_{q+1}, \ldots, J_n].$$

Since the pair $[C, D]$ is nondegenerate, there are n linearly independent vectors among its $2n$ columns. They can be taken to be the last $n - q$ columns $J_{q+1}, \ldots, J_n$ of D and some q columns $K_{j_1}, \ldots, K_{j_q}$ of C.

Then the matrix $F = [K_{j_1}, \ldots, K_{j_q}, J_{q+1}, \ldots, J_n]$ is such that $\det F \neq 0$.

It will be assumed that $j_1 < j_2 < \cdots < j_q$; let $j_s \leq q < j_{s+1}$. The elements of the set

$$\{1, \ldots, q\} \backslash \{j_1, \ldots, j_s\} = \{g_1, \ldots, g_{q-s}\}$$

are numbered in order of increase.

We arrange the columns of the matrix C by multiplying it from the right by the symmetric matrix Z_0 in such a way that the columns $K_{j_1}, \ldots, K_{j_q}$ of C become the first q columns of CZ_0.

For this it suffices to assign the values of the elements z_{ik} of Z_0 as follows:

1) if $k = j_{s+t}$ $(t = 1, \ldots, q - s)$, then

$$z_{ik} = \begin{cases} 1 & \text{for } i = g_t, \\ 0 & \text{for } i \neq g_t; \end{cases}$$

2) if $k = g_t$ $(t = 1, \ldots, q - s)$, then

$$z_{ik} = \begin{cases} 1 & \text{for } i = j_{s+t}, \\ 0 & \text{for } i \neq j_{s+t}; \end{cases}$$

3) if $k \notin \{j_{s+1}, \ldots, j_q\} \cup \{g_1, \ldots, g_t\}$, then

$$z_{ik} = \begin{cases} 1 & \text{for } i = k, \\ 0 & \text{for } i \neq k. \end{cases}$$

Let $Z = \lambda Z_0$, where λ is an arbitrary real scalar. The matrix $CZ + D$ has the form

$$C\lambda Z_0 + D = [\lambda K_{l_1}, \ldots, \lambda K_{l_q}, \lambda K_{l_{q+1}} + J_{q+1}, \ldots, \lambda K_{l_n} + J_n],$$

where the columns $K_{l_1}, \ldots, K_{l_q}$ coincide in some order with $K_{j_1}, \ldots, K_{j_q}$.

If in the matrix $C\lambda Z_0 + D$ we divide the first columns $\lambda K_{l_1}, \ldots, \lambda K_{l_q}$ by λ and pass to the limit as $\lambda \to 0$, then we get a matrix G differing from F at most by the order of the first q columns, and hence $\det G \neq 0$. On the other hand, since $\det(CZ + D) \equiv 0$ by assumption, it follows that $\det G = \lim(1/\lambda^2) \det(C\lambda Z_0 + D) = 0$. This contradiction shows that the statement of the theorem is valid.

It is sometimes necessary to consider a linear fractional transformation of the pair $[p, q]$:

$$y = (ap + bq)(cp + dq)^{-1}.$$

Here the linear fractional transformation (1) is a special case of a transformation of a pair.

The transition to pairs is analogous to the introduction of projective coordinates, and enables us to differentiate various cases of "infinitization" of the matrix x.

General part

Essential for what follows is the fact that any linear fractional transformation of the form (1) (with nonsingular matrix A) can also be written in the form (2), and conversely.

THEOREM 2. *The transformations* (1) *and* (2) *coincide if and only if their matrices*

$$A = \begin{pmatrix} a & b \\ c & d \end{pmatrix}, \qquad B = \begin{pmatrix} a_1 & c_1 \\ b_1 & d_1 \end{pmatrix}$$

are connected by the relation

$$B \begin{pmatrix} 0 & I \\ -I & 0 \end{pmatrix} A = \rho \begin{pmatrix} 0 & I \\ -I & 0 \end{pmatrix}, \tag{3}$$

where ρ is a scalar (nonzero).

PROOF. Suppose that (1) and (2) coincide. Then for any matrix x

$$\begin{aligned}
0 &= (xc_1 + d_1)^{-1}(xa_1 + b_1) - (ax + b)(cx + d)^{-1} \\
&= (xc_1 + d_1)^{-1}\{(xa_1 + b_1)(cx + d) - (xc_1 + d_1)(ax + b)\}(cx + d)^{-1},
\end{aligned}$$

or

$$x(a_1 c - c_1 d)x + x(a_1 a - c_1 b) + (d_1 a - b_1 c)x + (b_1 d - d_1 b) = 0. \tag{4}$$

Setting $x = tI$ here, where t is a real scalar, and letting t go to 0 and ∞, we get that

$$a_1 c - c_1 a = 0, \qquad b_1 d - d_1 b = 0, \tag{5}$$

after which (4) takes the form $x(a_1 d - c_1 b) = (d_1 a - b_1 c)x$. Setting $x = I$ here, we get that

$$a_1 d - c_1 b = d_1 a - b_1 c = l, \tag{5'}$$

$$xl = lx. \tag{6}$$

Assuming that x is a diagonal real matrix with distinct diagonal elements, we get from (6) that l is also diagonal:

$$l = \begin{pmatrix} \rho_1 & 0 & \cdots & 0 \\ 0 & \rho_2 & \cdots & 0 \\ \cdots & \cdots & \cdots & \cdots \\ 0 & 0 & \cdots & \rho_n \end{pmatrix}.$$

Finally, taking x to be a matrix of the form

$$x = \begin{pmatrix} 0 & 1 & 0 & & \\ 1 & 0 & 1 & & \\ 0 & 1 & 0 & \ddots & \\ & & \ddots & \ddots & 1 \\ & & & 1 & 0 \end{pmatrix}$$

we see that the diagonal elements of l are the same, i.e., $l = \rho I$ ($\rho \neq 0$). But then (5) and (5') are equivalent to (3).

Suppose, conversely, that A and B are connected by relation (3). Then (4) holds, and hence the transformations (1) and (2) coincide.

The proof of Theorem 2 has an important corollary.

COROLLARY. *The transformations* (1) *and* (2) *coincide if the values of their right-hand sides coincide on real symmetric matrices.*

With each linear fractional transformation of the form (1) in the class under consideration we associate its coefficient matrix A. We show that this correspondence uniquely determines A to within a scalar factor.

THEOREM 3. *The linear fractional transformations*

$$y = (ax + b)(cx + d)^{-1}, \tag{7}$$

$$y = (a_1 x + b_1)(c_1 x + d_1)^{-1} \tag{8}$$

coincide if and only if their coefficient matrices are collinear, i.e., if and only if

$$A_1 = \rho A \qquad (\rho \neq 0). \tag{9}$$

PROOF. Suppose that (7) and (8) are the same transformation, and let it be written in the form (2). Then

$$B \begin{pmatrix} 0 & I \\ -I & 0 \end{pmatrix} A = \rho_0 \begin{pmatrix} 0 & I \\ -I & 0 \end{pmatrix}, \qquad B \begin{pmatrix} 0 & I \\ -I & 0 \end{pmatrix} A_1 = \rho_1 \begin{pmatrix} 0 & I \\ -I & 0 \end{pmatrix},$$

which implies (9) with $\rho = \rho_1/\rho_0$.

The converse assertion is trivial.

COROLLARY. *The transformations* (7) *and* (8) *coincide if the values of their right-hand sides coincide on real symmetric matrices.*

Indeed, write (7) in the form (2) and use the corollary to Theorem 2.

Let us dwell on the group properties of linear fractional transformations (1) with nonsingular coefficient matrix A.

First of all, we consider the composition of two linear fractional transformations

$$y = (a_1 x + b_1)(c_1 x + d_1)^{-1}, \qquad z = (a_2 y + b_2)(c_2 y + d_2)^{-1}. \tag{10}$$

We get that

$$\begin{aligned} z &= [a_2(a_1 x + b_1)(c_1 x + d_1)^{-1} + b_2][c_2(a_1 x + b_1)(c_1 x + d_1)^{-1} + d_2]^{-1} \\ &= [a_2(a_1 x + b_1) + b_2(c_1 x + d_1)][c_2(a_1 x + b_1) + d_2(c_1 x + d_1)]^{-1} \\ &= [(a_2 a_1 + b_2 c_1)x + (a_2 b_1 + b_2 d_1)][(c_2 a_1 + d_2 c_1)x + (c_2 b_1 + d_2 d_1)]^{-1}. \end{aligned}$$

Thus, the matrix A_3 of the composition $z = (a_3 x + b_3)(c_3 x + d_3)^{-1}$ is collinear with the product $A_2 A_1$, i.e., $A_3 = \rho A_2 A_1$.

Further, suppose that the transformations in (10) are mutually inverse, i.e., the composition of them is the identity transformation

$$Z = X = (Ix + 0)(0x + I)^{-1}.$$

The matrix A_3 of this transformation is collinear with the identity; therefore A_2 is collinear with the inverse A_1^{-1}. The same result can be obtained in another way. For the transformation $y = (a_1 x + b_1)(c_1 x + d_1)^{-1}$ we find the inverse

$$x = (-yc_1 + a_1)^{-1}(yd_1 - b_1).$$

Corresponding to it is the matrix

$$B_1 = \begin{pmatrix} d_1 & -c_1 \\ -b_1 & a_1 \end{pmatrix}.$$

The inverse transformation can be written in the form (1):

$$x = (a_2 y + b_2)(c_2 y + d_2)^{-1};$$

then, by (3),

$$B_1 \begin{pmatrix} 0 & I \\ -I & 0 \end{pmatrix} A_2 = \rho \begin{pmatrix} 0 & I \\ -I & 0 \end{pmatrix},$$

but since clearly

$$\begin{pmatrix} 0 & -I \\ I & 0 \end{pmatrix} B_1 \begin{pmatrix} 0 & I \\ -I & 0 \end{pmatrix} = A_1,$$

it follows that $A_1 A_2 = \rho I$.

This argument is given here because the use of (3) is a distinctive stamp in the proofs of a number of the theorems to follow.

THEOREM 4. *The linear fractional transformation* (1) *carries symmetric matrices x into symmetric matrices y if and only if A is collinear to a symplectic matrix, i.e.,*

$$A' \begin{pmatrix} 0 & I \\ -I & 0 \end{pmatrix} A = \rho^2 \begin{pmatrix} 0 & I \\ -I & 0 \end{pmatrix}. \tag{11}$$

PROOF. Suppose that (1) carries $x = x'$ into $y = y'$. Then the transformations

$$y = (ax + b)(cx + d)^{-1}, \tag{12}$$
$$y = (xc' + d')^{-1}(xa' + b') \tag{13}$$

coincide on symmetric matrices x. By the corollary to Theorem 1, they coincide identically, and then (3) holds by Theorem 2, which gives us (11).

Suppose, conversely, that (11) holds. Then, by Theorem 2, the transformations (12) and (13) coincide, and we get that for a symmetric matrix

$$y' = \{(ax + b)(cx + d)^{-1}\}' = (xc' + d')^{-1}(xa' + b') = y.$$

COROLLARY. *The linear fractional transformation* (1) *carries transposed matrices x' into transposed matrices y', i.e.,*

$$y' = (ax' + b)(cx' + d)^{-1}, \tag{14}$$

provided that it carries symmetric matrices x into symmetric matrices y.

Indeed, (13) coincides with (12). Transposing (13), we get (14).

The proof of the next result is completely analogous.

THEOREM 5. *The linear fractional transformation* (1) *carries Hermitian matrices x into Hermitian matrices y if and only if A is collinear with the unitary* *matrix* $\begin{pmatrix} 0 & iI \\ -iI & 0 \end{pmatrix}$, *i.e.*,

$$A^* \begin{pmatrix} 0 & I \\ -I & 0 \end{pmatrix} A = \rho^2 \begin{pmatrix} 0 & I \\ -I & 0 \end{pmatrix}.$$

Here ρ^2 is a real number, and ρ can be purely imaginary.

COROLLARY. *The linear fractional transformation* (1) *carries conjugate matrices x^* into conjugate matrices y^* if it carries Hermitian matrices x into Hermitian matrices y.*

The proof of Theorem 6 is almost the same as that of Theorem 4.

THEOREM 6. *The linear fractional transformation* (1) *carries real matrices x into real matrices if and only if A is collinear with a real matrix, i.e., $A = \rho A_0$, where $\overline{A}_0 = A_0$.*

PROOF. Suppose that (1) carries $x = \bar{x}$ into $y = \bar{y}$. Then the transformations

$$y = (ax + b)(cx + d)^{-1}, \tag{15}$$

$$y = (\bar{a}x + \bar{b})(\bar{c}x + \bar{d})^{-1} \tag{16}$$

coincide on the real matrices x. Writing (15) in the form (2), we get the equalities

$$B \begin{pmatrix} 0 & I \\ -I & 0 \end{pmatrix} A = \rho_0 \begin{pmatrix} 0 & I \\ -I & 0 \end{pmatrix}, \qquad B \begin{pmatrix} 0 & I \\ -I & 0 \end{pmatrix} \overline{A} = \rho_1 \begin{pmatrix} 0 & I \\ -I & 0 \end{pmatrix},$$

which imply that $\overline{A} = e^{i\alpha} A$, and, consequently, $A_0 = e^{i\alpha/2} A$ is a real matrix. The converse assertion is trivial.

COROLLARY. *The linear fractional transformation* (1) *carries complex conjugate matrices $\bar{x}$ into complex conjugate matrices $\bar{y}$ if it carries real matrices x into real matrices y.*

THEOREM 7. *If a symplectic matrix A is collinear with a real matrix, then it is either real or purely imaginary.*

Indeed, since

$$A' \begin{pmatrix} 0 & I \\ -I & 0 \end{pmatrix} A = \begin{pmatrix} 0 & I \\ -I & 0 \end{pmatrix}, \qquad A = \rho A_0, \quad \overline{A}_0 = A_0,$$

it follows that

$$\rho^2 A'_0 \begin{pmatrix} 0 & I \\ -I & 0 \end{pmatrix} A_0 = \begin{pmatrix} 0 & I \\ -I & 0 \end{pmatrix},$$

and since ρ^2 is a real number, ρ is either real or purely imaginary.

It follows from Theorem 7 that a linear fractional transformation (1) carrying real matrices into real matrices and symmetric matrices into symmetric matrices can always be normalized in such a way that A is a real symplectic matrix or a real antisymplectic matrix; the latter means that

$$A' \begin{pmatrix} 0 & I \\ -I & 0 \end{pmatrix} A = - \begin{pmatrix} 0 & I \\ -I & 0 \end{pmatrix}.$$

If the matrix of the transformation

$$y = (ax + b)(cx + d)^{-1}$$

is symplectic, then the matrix of the transformation

$$z = (cx + d)(ax + b)^{-1}$$

is antisymplectic.

Special part

We consider linear fractional transformations (1) carrying the right matrix half-plane (i.e., the collection of matrices x such that $x^* + x \geq 0$) into itself. Transformations (1) carrying the J-disk into the J-disk, the J-disk into a half-plane, and so on, can be studied in a similar way.

Suppose that an indefinite metric is given in a finite-dimensional complex linear space L by the innner product $[f, g]$.

A vector $f \in L$ is said to be *positive* if $[f, f] > 0$, *negative* if $[f, f] < 0$, and *neutral* if $[f, f] = 0$. For example, for $2n$-dimensional column vectors we can let $[f, g] = fJg^*$, where $J = \begin{pmatrix} 0 & I \\ I & 0 \end{pmatrix}$. The matrix A is a *plus-matrix* if it carries each nonnegative vector f into a nonnegative vector, i.e., if the condition $[f, f] \geq 0$ implies that $[fA, fA] \geq 0$, or, in the concrete realization under consideration, if the condition

$$fJf^* \geq 0 \tag{17}$$

implies that

$$fAJA^*f^* \geq 0. \tag{18}$$

It will be shown that if the nonsingular matrix A is a plus-matrix, then so is A^*, and conversely.

THEOREM 8. *The linear fractional transformation* (1) *carries the right matrix half-plane* $x^* + x \geq 0$ *into itself if and only if the coefficient matrix is a plus-matrix.*

PROOF. We compute

$$\begin{aligned}
y^* + y &= (x^*c^* + d^*)^{-1}(x^*a^* + b^*) + (ax + b)(cx + d)^{-1} \\
&= (x^*c^* + d^*)^{-1}\{(x^*a^* + b^*)(cx + d) + (x^*c^* + d^*)(ax + b)\}(cx + d)^{-1} \\
&= (x^*c^* + d^*)^{-1}\{x^*(a^*c + c^*a)x + x^*(a^*d + c^*b) \\
&\qquad\qquad + (b^*c + d^*a)x + (b^*d + d^*b)\}(cx + d)^{-1};
\end{aligned}$$

but

$$[x^*, I]J \begin{bmatrix} x \\ I \end{bmatrix} = [x^*, I]\begin{pmatrix} 0 & I \\ I & 0 \end{pmatrix}\begin{bmatrix} x \\ I \end{bmatrix} = x^* + x,$$

$$\begin{aligned}
[x^*, I]A^*JA \begin{bmatrix} x \\ I \end{bmatrix} &= [x^*, I]\begin{pmatrix} c^*a + a^*c & c^*b + a^*d \\ d^*a + b^*c & d^*b + b^*d \end{pmatrix}\begin{bmatrix} x \\ I \end{bmatrix} \\
&= x^*(c^*a + a^*c)x + x^*(a^*d + c^*b) \\
&\qquad + (b^*c + d^*a)c + (b^*d + d^*b).
\end{aligned}$$

Therefore,

$$y^* + y = (x^*c^* + d^*)^{-1}\left\{[x^*, I]A^*JA\begin{bmatrix}x\\I\end{bmatrix}\right\}(cx+d)^{-1},$$

and the condition $x^* + x \geq 0$ implies that $y^* + y \geq 0$ if and only if the condition

$$[x^*, I]J\begin{bmatrix}x\\I\end{bmatrix} \geq 0 \tag{19}$$

implies that

$$[x^*, I]A^*JA\begin{bmatrix}x\\I\end{bmatrix} \geq 0. \tag{20}$$

It is not hard to see that the last condition is equivalent to A^* being a plus-matrix.

Indeed, suppose for the matrix A that (19) implies (20) and assume first that $[u, v]$ is a proper pair of matrices (i.e., v^{-1} exists) satisfying the condition

$$[u, v]J\begin{bmatrix}u^*\\v^*\end{bmatrix} \geq 0. \tag{21}$$

We show that $[u, v]$ also satisfies the condition

$$[u, v]A^*JA\begin{bmatrix}u^*\\v^*\end{bmatrix} \geq 0. \tag{22}$$

Setting $v^{-1}u = x^*$, we get that $[u, v] = v[x^*, I]$, and, since the pair $[u, v]$ satisfies (21), x satisfies (19). But (19) implies (20), and (20) implies (22).

Suppose now that $[u, v]$ is an arbitrary pair satisfying (21). We show that $[u, v]$ can be regarded as a limit of proper pairs also satisfying (21).

With this goal we consider a nonsingular matrix T having the property that $Tu^* = uu^*$.

For $\sigma > 0$ the pair $[u, v+\sigma T]$ satisfies condition (21) when $[u, v]$ does, because

$$[u, v + \sigma T]J\begin{bmatrix}u^*\\v^* + \sigma T^*\end{bmatrix} = u(v^* + \sigma T^*) + (v + \sigma T)u^* = uv^* + vu^* + 2\sigma uu^* \geq 0.$$

On the other hand, for all sufficiently small σ the matrix $v+\sigma T$ is nonsingular, and, by what was just proved, the pair $[u, v + \sigma T]$ satisfies (22). Passing to the limit as $\sigma \to 0$, we get that $[u, v]$ also satisfies (22).

Finally, suppose that f is an arbitrary $2n$-dimensional vector satisfying

$$fJf^* \geq 0. \tag{17'}$$

Let $f = (\xi_1, \ldots, \xi_n, \eta_1, \ldots, \eta_n) = (g_1, g_2)$; then

$$fJf^* = (g_1, g_2)\begin{pmatrix}0 & I\\I & 0\end{pmatrix}\begin{pmatrix}g_1^*\\g_2^*\end{pmatrix} = g_1g_2^* + g_2g_1^* \geq 0.$$

We show that

$$fA^*JAf^* \geq 0. \tag{18'}$$

Consider the pair $[u, v]$ of $n \times n$ matrices with

$$u = \begin{pmatrix} \xi_1 & \xi_2 & \cdots & \xi_n \\ 0 & 0 & \cdots & 0 \\ \cdots & \cdots & \cdots & \cdots \\ 0 & 0 & \cdots & 0 \end{pmatrix}, \qquad v = \begin{pmatrix} \eta_1 & \eta_2 & \cdots & \eta_n \\ 0 & 0 & \cdots & 0 \\ \cdots & \cdots & \cdots & \cdots \\ 0 & 0 & \cdots & 0 \end{pmatrix}.$$

This pair satisfies (21), since

$$[u, v]J \begin{bmatrix} u^* \\ v^* \end{bmatrix} = \begin{pmatrix} g_2 g_1^* + g_1 g_2^* & 0 & \cdots & 0 \\ 0 & 0 & \cdots & 0 \\ \cdots & \cdots & \cdots & \cdots \\ 0 & 0 & \cdots & 0 \end{pmatrix} \geq 0.$$

But then $[u, v]$ satisfies (22),

$$[u, v]A^* J A \begin{bmatrix} u^* \\ v^* \end{bmatrix} = \begin{pmatrix} f A^* J A f^* & 0 & \cdots & 0 \\ 0 & 0 & \cdots & 0 \\ \cdots & \cdots & \cdots & \cdots \\ 0 & 0 & \cdots & 0 \end{pmatrix},$$

and this means that $(18')$ holds.

Conversely, suppose that A^* is a plus-matrix, and consider an arbitrary matrix x satisfying (19). Then for an arbitrary n-dimensional vector g

$$g[x^*, I]J \begin{bmatrix} x \\ I \end{bmatrix} g^* \geq 0$$

and the $2n$-dimensional vector $f = (gx^*, g)$ satisfies $(17')$, hence $(18')$, which now takes the form

$$g[x^*, I]A^* J A \begin{bmatrix} x \\ I \end{bmatrix} g^* \geq 0.$$

Since this inequality holds for any vector g, it follows that

$$[x^*, I]A^* J A \begin{bmatrix} x \\ I \end{bmatrix} \geq 0,$$

i.e., (19) implies (20). The theorem is proved.

As L. A. Sakhnovich first showed, each nonsingular plus-matrix is collinear with a J-expansive matrix, i.e., $A = \rho A_0$, where $A_0 J A_0^* - J \geq 0$.

This gives us, in particular, that a transformation carrying the right matrix half-plane into itself makes sense in the whole open right half-plane. Namely, we have

THEOREM 9. *If A^* is a nonsingular J-expansive matrix, then the linear fractional transformations*

$$y = (ax + b)(cx + d)^{-1}, \qquad y = (cx + d)(ax + b)^{-1}$$

make sense for each matrix x such that $x^* + x > 0$.

PROOF. By assumption,

$$[x^*, I]A^* J A \begin{bmatrix} x \\ I \end{bmatrix} \geq [x^*, I]J \begin{bmatrix} x \\ I \end{bmatrix}.$$

However,

$$[x^*, I]\begin{pmatrix} 0 & I \\ I & 0 \end{pmatrix}\begin{bmatrix} x \\ I \end{bmatrix} = x^* + x,$$

$$[x^*, I]\begin{pmatrix} a^* & c^* \\ b^* & d^* \end{pmatrix}\begin{pmatrix} 0 & I \\ I & 0 \end{pmatrix}\begin{pmatrix} a & b \\ c & d \end{pmatrix}\begin{bmatrix} x \\ I \end{bmatrix}$$

$$= [x^*a^* + b^*, x^*c^* + d^*]\begin{pmatrix} 0 & I \\ I & 0 \end{pmatrix}\begin{bmatrix} ax + b \\ cx + d \end{bmatrix}$$

$$= (x^*c^* + d^*)(ax + b)(x^*a^* + b^*)(cx + d).$$

Thus,

$$(x^*c^* + d^*)(ax + b) + (x^*a^* + b^*)(cx + d) \geq x^* + x > 0$$

and if $cx + d$ is singular, then there exists a vector $f \neq 0$ such that $(cx+d)f^* = 0$. But then

$$0 = f\{(x^*c^* + d^*)(ax + b) + (x^*a^* + b^*)(cx + d)\}f^* \geq f(x^* + x)f^* > 0,$$

which is impossible.

It is proved similarly that $ax + b$ is a nonsingular matrix. The theorem is proved.

To establish that a plus-matrix is collinear with a J-expansive matrix we first prove a theorem of Kühne.

THEOREM 10. *If a Hermitian bilinear form* $\Omega(f, g)$ *is nonnegative on all neutral vectors* f, *i.e., if the condition* $[f, f] = 0$ *implies that* $\Omega(f, f) \geq 0$, *then for any positive vector* x *and any negative vector* y

$$\Omega(x, x)/[x, x] \geq \Omega(y, y)/[y, y] \tag{23}$$

and there exists a scalar μ *such that for all vectors* f

$$\Omega(f, f) \geq \mu[f, f]. \tag{24}$$

PROOF. Assume that (23) does not hold. Then there exist vectors x_0 and y_0 for which the opposite inequality holds. Normalizing them so that $[x_0, x_0] = 1$ and $[y_0, y_0] = -1$, we get that

$$\Omega(x_0, x_0) + \Omega(y_0, y_0) < 0. \tag{25}$$

Now let $f = x_0 + e^{i\alpha}y_0$, and choose the scalar factor $e^{i\alpha}$ such that the vector f is neutral, i.e., so that

$$[f, f] = [x_0 + e^{i\alpha}y_0, x_0 + e^{i\alpha}y_0]$$

$$= [x_0, x_0] + e^{-i\alpha}[x_0, y_0] + e^{i\alpha}[y_0, x_0] + [y_0, y_0]$$

$$= e^{-i\alpha}[x_0, y_0] + e^{i\alpha}[y_0, x_0]$$

$$= \mathrm{Re}\{e^{-i\alpha}[x_0, y_0]\} = 0.$$

Obviously, there always exist two such factors $\pm e^{i\theta}$ of opposite signs. Let $f_1 = x_0 + e^{i\theta}y_0$ and $f_2 = x_0 - e^{i\theta}y_0$ be two corresponding vectors. Since one of the two real numbers

$$\pm[e^{-i\theta}\Omega(x_0, y_0) + e^{i\theta}\Omega(y_0, x_0)]$$

is automatically nonpositive, one of the two forms $\Omega(f_1, f_1)$ and $\Omega(f_2, f_2)$ must be negative, and this contradicts a condition of the theorem. Thus inequality (23) is proved.

Now let

$$\inf_{[x,x]>0} \frac{\Omega(x,x)}{[x,x]} = \mu. \tag{26}$$

By (23), we have that $\mu > -\infty$, and, for all vectors f, $\Omega(f,f) \geq \mu[f,f]$. Indeed, for positive vectors this follows from (26), for netural vectors it follows from the definition of $\Omega(f,g)$, and for negative vectors it follows from an inequality that is a consequence of (23):

$$\mu \geq \Omega(y,y)/[y,y].$$

(Here the inequality sign reverses upon multiplication of both sides of the inequality by the negative number $[y,y]$.)

The theorem is proved.

THEOREM 11. *Each nonsingular plus-matrix A is collinear to a J-noncontractive matrix.*

PROOF. The bilinear form $\Omega(f,g) = [fA, gA]$ satisfies the conditions of the preceding theorem. Therefore, $[fA, fA] \geq \mu[f,f]$ for nonnegative μ. If $\mu = 0$, then A maps the whole space into a proper subset, namely, the nonnegative subspace, and is thus singular. Hence, $\mu = 1/\rho^2 > 0$, and the matrix ρA is J-noncontractive.

THEOREM 12. *If A is a J-noncontractive matrix, then so is A^*, and conversely. Thus, the relations $AJA^* - J \geq 0$ and $A^*JA - J \geq 0$ are equivalent.*

PROOF. Consider the Cayley transformation

$$w = J(I + e^{i\theta}u)(I - e^{i\theta}u)^{-1}, \tag{27}$$

where the scalar factor $e^{i\theta}$ is chosen so that the transformation makes sense for $u = A$. Then

$$\frac{w + w^*}{2} = (I - e^{-i\theta}u^*)^{-1}\{J - u^*Ju\}(I - e^{i\theta}u)^{-1}; \tag{28}$$

on the other hand, (27) can be written in the form

$$w = J(I - e^{i\theta}u)^{-1}(I + e^{i\theta}u), \tag{29}$$

and then

$$\frac{w + w^*}{2} = J(I - e^{i\theta}u)^{-1}\{J - uJu^*\}(I - e^{-i\theta}u^*)^{-1}J. \tag{30}$$

From (28) and (30) we get that for $u = A$

$$J - AJA^* = T(J - A^*JA)T^*,$$

where T is a nonsingular matrix, and this proves the theorem.

COROLLARY. *If A is a nonsingular plus-matrix, then so is A^*, and conversely.*

Indeed, if A is a nonsingular plus-matrix, then by Theorem 11 there exists a scalar ρ such that ρA is a J-noncontractive matrix. By Theorem 12, ρA^* is also J-noncontractive; in particular, A^* is a plus-matrix.

REMARK. The corollary does not hold for singular matrices. Let

$$J = \begin{pmatrix} 1 & 0 \\ 0 & -1 \end{pmatrix}, \qquad \varepsilon = \begin{pmatrix} 1 & -1 \\ 1 & -1 \end{pmatrix}, \qquad A = \begin{pmatrix} \alpha & 0 \\ 0 & 1 \end{pmatrix} \varepsilon;$$

then

$$AJA^* = \begin{pmatrix} \alpha & 0 \\ 0 & 1 \end{pmatrix} \varepsilon J \varepsilon^* \begin{pmatrix} \bar{\alpha} & 0 \\ 0 & 1 \end{pmatrix} = 0,$$

i.e., in particular, A is a plus-matrix. However,

$$A^*JA = \varepsilon^* \begin{pmatrix} |\alpha|^2 & 0 \\ 0 & 1 \end{pmatrix} \varepsilon = \varepsilon^* J \varepsilon + \varepsilon^* \begin{pmatrix} |\alpha|^2 - 1 & 0 \\ 0 & 0 \end{pmatrix} \varepsilon$$

$$= \varepsilon^* \begin{pmatrix} 1 & 0 \\ 0 & 0 \end{pmatrix} \varepsilon(|\alpha|^2 - 1),$$

and A^* is not a plus-matrix for $|\alpha| < 1$.

It will again be assumed that the matrix J has the form $J = \begin{pmatrix} 0 & I \\ I & 0 \end{pmatrix}$. Moreover, we introduce the matrix

$$j = \begin{pmatrix} -I & 0 \\ 0 & I \end{pmatrix}.$$

Then the symplectic property of the matrix can obviously be expressed as

$$A'JjA = Jj. \tag{31}$$

THEOREM 13. *A symplectic plus-matrix A is a J-noncontractive matrix.*

PROOF. By Theorem 11, there exists a scalar $\rho > 0$ such that the matrix ρA is J-noncontractive, i.e.,

$$\rho^2 AJA^* > J. \tag{32}$$

By (31), $A = jJA'^{-1}Jj$ and $A^* = jJ\bar{A}^{-1}Jj$. Substituting in (32), we get that

$$\rho^2 jJA'^{-1}JjJjJ\bar{A}^{-1}Jj \geq J$$

or $-\rho^2 A'^{-1}J\bar{A}^{-1} \geq J$, and hence $A'J\bar{A} \geq \rho^2 J$. We proceed to the complex conjugate: $A^*JA \geq \rho^2 J$. However, by Theorem 12, this inequality is equivalent to

$$AJA^* \geq \rho^2 J. \tag{33}$$

Adding (32) and (33), we get that

$$(1 + \rho^2)AJA^* \geq (1 + \rho^2)J$$

or $AJA^* \geq J$, which is what we wanted to prove.

The following basic result follows from Theorems 7 and 13.

THEOREM 14. *If the linear fractional transformation*

$$y = (ax + b)(cx + d)^{-1}$$

with nonsingular coefficient matrix $A = \begin{pmatrix} a & b \\ c & d \end{pmatrix}$ *maps the right matrix half-plane* $x^* + x \geq 0$ *into itself and carries real symmetric matrices* x *into real symmetric matrices* y, *then by multiplying all the coefficients by a scalar factor* ρ *the transformation can be normalized in such a way that the matrix* ρA *is a real* J-*noncontractive symplectic or antisymplectic matrix.*

BIBLIOGRAPHY

1. V. P. Potapov, *The multiplicative structure of J-nonexpansive matrix-valued functions*, Trudy Moskov. Mat. Obshch. **4** (1955), 125–236; English transl. in Amer. Math. Soc. Transl. (2) **15** (1960).

2. ______, *General theorems on the structure and the splitting of elementary factors of analytic matrix-valued functions*, Akad. Nauk Armyan. SSR Dokl. **48** (1969), 257–263; English transl. in this volume.

3. A. V. Efimov and V. P. Potapov, *J-expansive matrix-valued functions and their role in the analytic theory of electric circuits*, Uspekhi Mat. Nauk **28** (1973), no. 1 (169), 65–130; English transl. in Russian Math. Surveys **28** (1973).

4. I. V. Kovalishina and V. P. Potapov, *An indefinite metric in the Nevanlinna-Pick problem*, Akad. Nauk Armyan. SSR Dokl. **59** (1974), 17–22; English transl. in this volume.

5. I. V. Kovalishina, *J-expansive matrix-valued functions in the Carathéodory problem*, Akad. Nauk Armyan. SSR Dokl. **59** (1974), 129–135. (Russian)

6. ______, *J-expansive matrix-valued functions and the classical moment problem*, Akad. Nauk Armyan. SSR Dokl. **60** (1975), 3–10. (Russian)

7. ______, *Additive decomposition of an arbitrary reactive matrix-valued function*, Izv. Akad. Nauk Armyan. SSR Ser. Mat. **6** (1971), 43–60. (Russian)

8. Carl Ludwig Siegel, *Symplectic geometry*, Amer. J. Math. **65** (1943), 1–86.

9. Hua Loo-Keng [Hua Luo-geng], *Geometries of matrices. Parts I, I$_1$, II, III*, Trans. Amer. Math. Soc. **57** (1945), 441–481, 482–490; **61** (1947), 193–228, 229–255.

Translated by H. H. MCFADEN

Amer. Math. Soc. Transl.
(2) Vol. **138**, 1988

The Radii of a Weyl Disk
in the Matrix Nevanlinna-Pick Problem

UDC 519.210

I. V. KOVALISHINA AND V. P. POTAPOV

In this article we consider additional facts about the matrix Nevanlinna-Pick problem presented previously by the authors [3]. The main result here is a derivation of multiplicative formulas for the radii of Weyl disks. On the basis of this a method is presented for constructing Nevanlinna-Pick problems with any previously specified ranks of the limit Weyl radii. The article opens the possibility of resolving analogous questions for other problems in which the situation is considered, in particular for selfadjoint differential equations with complex coefficients.

We derive formulas for the left and right radii of a Weyl disk, which play an essential role in the classification of Nevanlinna-Pick problems, by using the multiplicative structure of the matrix-valued function $\mathfrak{A}_k = f_k f_{k-1} \cdots f_2 f_1$ that is the coefficient matrix of the linear fractional transformation determining the general solution of the kth truncated problem. It will be assumed that the independent variable ς varies in the unit disk.

§1. It is natural to begin the investigation with an analysis of a binomial factor. Here we find the radii of the Weyl disk for a binomial factor $f^{-1}(\varsigma)$ and, conversely, we reproduce the binomial factor from the radii of the Weyl disk given at a particular point of the unit disk.

Since the algorithm for constructing the structural formulas necessarily leads us to the metric with $j = \begin{bmatrix} -I & 0 \\ 0 & I \end{bmatrix}$ already at the second step, the problem posed for a binomial factor is solved both in the metric $J = \begin{bmatrix} 0 & I \\ I & 0 \end{bmatrix}$ and in the metric $j = \begin{bmatrix} -I & 0 \\ 0 & I \end{bmatrix}$.

1980 *Mathematics Subject Classification* (1985 *Revision*). Primary 15A99, 30D50; Secondary 30E05, 47A99.

Translation of Theory of Operators in Function Spaces and its Applications (V. A. Marchenko, editor), "Naukova Dumka", Kiev, 1981, pp. 25–49; MR **84j**: 30050.

A binomial elementary factor, normalized to the modulus at the point $\varsigma = 0$, has the form

$$f^{-1}(\varsigma) = I - \left(1 - \frac{\varsigma_0 - \varsigma}{1 - \bar{\varsigma}_0 \varsigma} \cdot \frac{|\varsigma_0|}{\varsigma_0}\right) P,$$

where P is a projection of full rank which can be represented in the case $J = \begin{bmatrix} 0 & I \\ I & 0 \end{bmatrix}$ in the form

$$P = \begin{bmatrix} x^* \\ y^* \end{bmatrix} [x, y] J, \qquad [x, y] J \begin{bmatrix} x^* \\ y^* \end{bmatrix} = I,$$

and in the case $j = \begin{bmatrix} -I & 0 \\ 0 & I \end{bmatrix}$ in the form

$$P = \begin{bmatrix} \xi^* \\ \eta^* \end{bmatrix} [\xi, \eta] j, \qquad [\xi, \eta] j \begin{bmatrix} \xi^* \\ \eta^* \end{bmatrix} = I.$$

Here $x, y, \xi,$ and η are $m \times m$ matrices, with x and η nonsingular.

The Weyl matrix W will be considered at the point $\varsigma = 0$, which by assumption is not an interpolation node.

As before, in the J-metric we use the notation

$$W = \begin{bmatrix} -R & S^* \\ S & -T \end{bmatrix} = \begin{bmatrix} I & 0 \\ -SR^{-1} & I \end{bmatrix} \begin{bmatrix} -R & 0 \\ 0 & SR^{-1}S^* - T \end{bmatrix} \begin{bmatrix} I & -R^{-1}S^* \\ 0 & I \end{bmatrix}.$$

The corresponding lower-case letters of the Latin alphabet will be used for the j-metric.

THEOREM 1. *In the J-metric the radii of the Weyl disk corresponding to the binomial factor $f^{-1}(\xi)$ are determined from the formulas $\rho_d = R^{-1}$ and $\rho_g = |\varsigma_0|^2 R^{-1}$.*[1]

Only the second relation will be proved. Starting from the equality

$$PJ = \begin{bmatrix} x^*x & x^*y \\ y^*x & y^*y \end{bmatrix},$$

we compute the Weyl matrix

$$\begin{bmatrix} -R & S^* \\ S & -T \end{bmatrix} = f^{-1}(0) J f^{*-1}(0) = J - (1 - |\varsigma_0|^2) Py$$

$$= \begin{bmatrix} -(1 - |\varsigma_0|^2)x^*x & I - (1 - |\varsigma_0|^2)x^*y \\ I - (1 - |\varsigma_0|^2)y^*x & -(1 - |\varsigma_0|^2)y^*y \end{bmatrix}. \tag{1}$$

Then $R = (1 - |\varsigma_0|^2)x^*x$, and

$$\rho_g = SR^{-1}S^* - T$$

$$= [I - (1 - |\varsigma_0|^2)y^*x] \frac{x^{-1}x^{*-1}}{1 - |\varsigma_0|^2} [I - (1 - |\varsigma_0|^2)x^*y] - (1 - |\varsigma_0|^2)y^*y$$

$$= \frac{x^{-1}x^{*-1}}{1 - |\varsigma_0|^2} - x^{-1}[xy^* + yx^*]x^{*-1} = \frac{|\varsigma_0|^2}{1 - |\varsigma_0|^2} x^{-1}x^{*-1} = |\varsigma_0|^2 R^{-1},$$

which is what was required.

---

[1] d for *droit*, g for *gauche*.

The inverse problem is to construct a binomial factor from given radii $\rho_d = R^{-1}$ and $\rho_g = |\varsigma_0|^2 R^{-1}$. Obviously, its solution reduces to finding a projection

$$P = \begin{bmatrix} x^*x & x^*y \\ y^*x & y^*y \end{bmatrix} y, \qquad P^2 = P, \quad Py \geq 0.$$

The computations simplify if the Weyl matrix $W = \begin{bmatrix} -R & S^* \\ S & -T \end{bmatrix}$ is subjected to the transformation[2]

$$\begin{bmatrix} R^{-1/2} & 0 \\ 0 & R^{1/2} \end{bmatrix} W \begin{bmatrix} R^{-1/2} & 0 \\ 0 & R^{1/2} \end{bmatrix} = \begin{bmatrix} I & R^{-1/2}S^*R^{1/2} \\ R^{1/2}SR^{-1/2} & -R^{1/2}TR^{1/2} \end{bmatrix}.$$

Here the radii of the Weyl disk pass into $\rho_d = I$ and $\rho_g = |\varsigma_0|^2 I$. But since $R = (1 - |\varsigma_0|)^2 x^*x$, it follows that

$$x^*x = \frac{I}{1 - |\varsigma_0|^2}. \tag{2}$$

Further, transforming the relation

$$[x, y]J \begin{bmatrix} x^* \\ y^* \end{bmatrix} = xy^* + yx^* = I$$

by left multiplication by x^* and right multiplication by x, we get that $x^*xy^*x + x^*yx^*x = xy^* + yx^* = I$, or, after cancellation by $x^*x = (1 - |\varsigma_0|^2)^{-1}I$,

$$y^*x + x^*y = I, \tag{3}$$

which implies that

$$x^*y = \tfrac{1}{2}I + ih, \tag{4}$$

where h is an arbitrary Hermitian matrix.

Finally, since on the one hand $T = (1 - |\varsigma_0|^2)y^*y$, and on the other hand,

$$\begin{aligned} T &= SR^{-1}S^* - \rho_g = SR^{-1}S^* - |\varsigma_0|^2 R^{-1} \\ &= SS^* - |\varsigma_0|^2 I \\ &= [I - (1 - |\varsigma_0|^2)y^*x][I - (1 - |\varsigma_0|^2)x^*y] - |\varsigma_0|^2 I \\ &= [I - (1 - |\varsigma_0|^2)(\tfrac{1}{2}I - ih)][I - (1 - |\varsigma_0|^2)(\tfrac{1}{2}I + ih)] - |\varsigma_0|^2 I \\ &= (1 - |\varsigma_0|^2)^2(\tfrac{1}{4}I + h^2), \end{aligned}$$

it follows that

$$y^*y = (1 - |\varsigma_0|^2)(\tfrac{1}{4}I + h^2). \tag{5}$$

Thus,

$$P = \begin{bmatrix} x^*x & x^*y \\ y^*x & y^*y \end{bmatrix} J = \begin{bmatrix} x^*y & x^*x \\ y^*y & y^*x \end{bmatrix} = \begin{bmatrix} \tfrac{1}{2}I + ih & \dfrac{1}{1 - |\varsigma_0|^2}I \\ (1 - |\varsigma_0|^2)(\tfrac{1}{4}I + h^2) & \tfrac{1}{2}I - ih \end{bmatrix}. \tag{6}$$

---

[2] In circuit theory $\begin{bmatrix} R^{-1/2} & 0 \\ 0 & R^{1/2} \end{bmatrix}$ is the transfer matrix of an ideal transformer.

It is easy to see that, for an arbitrary Hermitian matrix h, P is a J-nonnegative projection, and that the Weyl matrix corresponding to the elementary factor

$$f^{-1}(\varsigma) = I - \left(1 - \frac{\varsigma_0 - \varsigma}{1 - \varsigma_0\varsigma}\frac{|\varsigma_0|}{\varsigma_0}\right)P,$$

has the form

$$W = \begin{bmatrix} -I & I - (1 - |\varsigma_0|^2)(\tfrac{1}{2}I + ih) \\ I - (1 - |\varsigma_0|^2)(\tfrac{1}{2}I - ih) & -(1 - |\varsigma_0|^2)^2(\tfrac{1}{4}I + h^2) \end{bmatrix}.$$

Returning to the radii originally given, we subject P and W to the inverse transformation mapping, which reduces to bracketing PJ and W from the left and from the right by the matrices $\begin{bmatrix} R^{1/2} & 0 \\ 0 & R^{-1/2} \end{bmatrix}$.

Thus, finally,

$$P = \begin{bmatrix} R^{1/2} & 0 \\ 0 & R^{-1/2} \end{bmatrix} \begin{bmatrix} \tfrac{1}{2}I + ih & \dfrac{1}{1 - |\varsigma_0|^2}I \\ (1 - |\varsigma_0|^2)(\tfrac{1}{4}I + h^2) & \tfrac{1}{2}I - ih \end{bmatrix} \begin{bmatrix} R^{-1/2} & 0 \\ 0 & R^{1/2} \end{bmatrix}$$

$$= \begin{bmatrix} \tfrac{1}{2}I + iR^{1/2}hR^{-1/2} & \dfrac{1}{1 - |\varsigma_0|^2}R \\ (1 - |\varsigma_0|^2)(\tfrac{1}{4}R^{-1} + R^{-1/2}h^2R^{-1/2}) & \tfrac{1}{2}I - iR^{-1/2}hR^{1/2} \end{bmatrix}; \tag{7}$$

$$W = \begin{bmatrix} -R & I - (1 - |\varsigma_0|^2)(\tfrac{1}{2}I + iR^{1/2}hR^{-1/2}) \\ I - (1 - |\varsigma_0|^2)(\tfrac{1}{2}I - iR^{-1/2}hR^{1/2}) & -(1 - |\varsigma_0|^2)^2(\tfrac{1}{4}R^{-1} + R^{-1/2}h^2R^{-1/2}) \end{bmatrix}, \tag{8}$$

which concludes the construction.

Consequently, to the given radii of the Weyl disk there corresponds an infinite set of binomial factors depending on the parameter h, which is an arbitrary Hermitian matrix.

In the j-metric the connection between the left and right radii, as well as the connection between the radii and the binomial factor, becomes considerably more complicated. In particular, the radii will be collinear only if an additional restriction is imposed on the projection P.

We first prove a lemma.

LEMMA 1. *For the radii of the Weyl disk corresponding to the elementary binomial factor*

$$f^{-1}(\varsigma) = I - \left(1 - \frac{\varsigma_0 - \varsigma}{1 - \varsigma_0\varsigma}\frac{|\varsigma_0|}{\varsigma_0}\right)p,$$

$$p = \begin{bmatrix} \xi^* \\ \eta^* \end{bmatrix}[\xi, \eta]j, \qquad [\xi, \eta]j\begin{bmatrix} \xi^* \\ \eta^* \end{bmatrix} = I,$$

to be determined in the j-metric by the formulas $\rho_d = r^{-1}$ and $\rho_g = |\varsigma_0|^2 r^{-1}$ where $r = I + (1 - |\varsigma_0|^2)\xi^*\xi$, it is necessary and sufficient that in addition

$$-\xi^*\xi + \eta^*\eta = I. \tag{9}$$

PROOF. The Weyl matrix here has the form

$$w = \begin{bmatrix} -r & s^* \\ s & -t \end{bmatrix} = b^{-1}(0)jb^{*-1}(0) = j - (1 - |\varsigma_0|^2)pj$$

$$= \begin{bmatrix} -I - (1 - |\varsigma_0|^2)\xi^*\xi & -(1 - |\varsigma_0|^2)\xi^*\eta \\ -(1 - |\varsigma_0|^2)\eta^*\xi & I - (1 - |\varsigma_0|^2)\eta^*\eta \end{bmatrix} \tag{10}$$

and, consequently, the right radius is equal to

$$\rho_d = r^{-1} = [I + (1 - |\varsigma_0|^2)\xi^*\xi]^{-1}.$$

We now transform the expression $\rho_g = sr^{-1}s^* - t$ for the left radius, using for the time being only the main relation for the projection

$$[\xi, \eta]j \begin{bmatrix} \xi^* \\ \eta^* \end{bmatrix} = -\xi\xi^* + \eta\eta^* = I. \tag{11}$$

From the easily verified identity

$$\xi[I + (1 - |\varsigma_0|^2)\xi^*\xi]^{-1} = [I + (1 - |\varsigma_0|^2)\xi\xi^*]^{-1}\xi,$$

we get that

$$\begin{aligned}
sr^{-1}s^* - t &= (1 - |\varsigma_0|^2)\eta^*\xi[I + (1 - |\varsigma_0|^2)\xi^*\xi]^{-1}(1 - |\varsigma_0|^2)\xi^*\eta \\
&\quad + I - (1 - |\varsigma_0|^2)\eta^*\eta \\
&= (1 - |\varsigma_0|^2)\eta^*[I + (1 - |\varsigma_0|^2)\xi\xi^*]^{-1}(1 - |\varsigma_0|^2)\xi\xi^*\eta \\
&\quad + I - (1 - |\varsigma_0|^2)\eta^*\eta \\
&= (1 - |\varsigma_0|^2)\eta^*\{I - [I + (1 - |\varsigma_0|^2)\xi\xi^*]^{-1}\}\eta \\
&\quad + I - (1 - |\varsigma_0|^2)\eta^*\eta \\
&= I - (1 - |\varsigma_0|^2)\eta^*[I + (1 - |\varsigma_0|^2)\xi\xi^*]^{-1}\eta \\
&= I - (1 - |\varsigma_0|^2)\eta^*[I + (1 - |\varsigma_0|^2)(\eta\eta^* - I)]^{-1}\eta \\
&= I - \left[I + \frac{|\varsigma_0|}{1 - |\varsigma_0|^2}\eta^{-1}\eta^{*-1}\right]^{-1} \\
&= \left[\frac{|\varsigma_0|^2}{1 - |\varsigma_0|^2}\eta^{-1}\eta^{*-1}\right]\left[I + \frac{|\varsigma_0|^2}{1 - |\varsigma_0|^2}\eta^{-1}\eta^{*-1}\right]^{-1} \\
&= |\varsigma_0|^2[(1 - |\varsigma_0|^2)\eta^*\eta + |\varsigma_0|^2 I]^{-1}.
\end{aligned}$$

Comparing the computed expression for ρ_g with $\rho_d = [I + (1 - |\varsigma_0|^2)\xi^*\xi]^{-1}$, we see that the equality $\rho_g = |\varsigma_0|^2\rho_d$ holds if and only if $-\xi^*\xi + \eta^*\eta = I$.

We solve also the inverse problem: construct a binomial factor satisfying the condition $-\xi^*\xi + \eta^*\eta = I$ from given radii $\rho_d = r^{-1}$ and $\rho_g = |\varsigma_0|^2 r^{-1}$ ($r^{-1} < I$).

As in the preceding case, this problem reduces to finding the projection

$$\overline{p} = \begin{bmatrix} \xi^*\xi & \xi^*\eta \\ \eta^*\xi & \eta^*\eta \end{bmatrix} j.$$

Since the Weyl matrix has the form

$$w = \begin{bmatrix} -r & s^* \\ s & -t \end{bmatrix} = \begin{bmatrix} -I - (1 - |\varsigma_0|^2)\xi^*\xi & -(1 - |\varsigma_0|^2)\xi^*\eta \\ -(1 - |\varsigma_0|^2)\eta^*\xi & I - (1 - |\varsigma_0|^2)\eta^*\eta \end{bmatrix},$$

it follows from the equality $r = I + (1 - |\varsigma_0|^2)\xi^*\xi$ that

$$\xi^*\xi = \frac{r - I}{1 - |\varsigma_0|^2}, \tag{12}$$

42   I. V. KOVALISHINA AND V. P. POTAPOV

which implies that

$$\xi = x \frac{\sqrt{r - I}}{\sqrt{1 - |\varsigma_0|^2}}, \tag{13}$$

where x is an arbitrary unitary matrix: $x^* x = I$.

Further, it follows from the condition $-\xi^* \xi + \eta^* \eta = I$ that

$$\eta^* \eta = I + \xi^* \xi = \frac{r - |\varsigma_0|^2 I}{1 - |\varsigma_0|^2}, \tag{14}$$

which gives us that

$$\eta = y \frac{\sqrt{r - |\varsigma_0|^2 I}}{\sqrt{1 - |\varsigma_0|^2}}, \qquad y^* y = I. \tag{15}$$

But then

$$\xi^* \eta = \frac{\sqrt{r - I}\, x^* y \sqrt{r - |\varsigma_0|^2 I}}{1 - |\varsigma_0|^2}$$

or, if $x^* y = z$, $z^* z = I$, then

$$\xi^* \eta = \frac{\sqrt{r - I}\, z \sqrt{r - |\varsigma_0|^2 I}}{1 - |\varsigma_0|^2}. \tag{16}$$

We now consider what condition must be satisfied by the tentatively arbitrary unitary matrix z. For this, in the main relation for p $(-\xi^* \xi + \eta^* \eta = I)$ we substitute the expressions (13) and (15) for ξ and η:

$$-x \frac{\sqrt{r - I}}{\sqrt{1 - |\varsigma_0|^2}} \cdot \frac{\sqrt{r - I}}{\sqrt{1 - |\varsigma_0|^2}} x^* + y \frac{\sqrt{r - |\varsigma_0|^2 I}}{\sqrt{1 - |\varsigma_0|^2}} \cdot \frac{\sqrt{r - |\varsigma_0|^2 I}}{\sqrt{1 - |\varsigma_0|^2}} y^* = I. \tag{17}$$

Multiplication of (17) on the left by x^* and on the right by y leads to

$$-(r - I)x^* y + x^* y(r - |\varsigma_0|^2 I) = (1 - |\varsigma_0|^2)x^* y,$$

or

$$rz = zr. \tag{18}$$

Finally, it is easy to verify that

$$p = \begin{bmatrix} \dfrac{r - I}{1 - |\varsigma_0|^2} & \dfrac{\sqrt{r - I}\, z \sqrt{r - |\varsigma_0|^2 I}}{1 - |\varsigma_0|^2} \\[4mm] \dfrac{\sqrt{r - |\varsigma_0|^2 I}\, z^* \sqrt{r - I}}{1 - |\varsigma_0|^2} & \dfrac{r - |\varsigma_0|^2 I}{1 - |\varsigma_0|^2} \end{bmatrix} j$$

is a j-nonnegative projection for any unitary matrix z commuting with r: we have $pj \geq 0$ and $p^2 = p$.

Thus, to the given radii of the Weyl disk there corresponds an infinite set of binomial factors such that in addition $-\xi^* \xi + \eta^* \eta = I$; this set depends on the parameter z, a unitary matrix commuting with r: $rz = zr$.

Note that the additional condition (9) is a consequence of the main relation $-\xi^* \xi + \eta^* \eta = I$ in the case when ξ and η are Hermitian or normal matrices.

Suppose now that

$$f^{-1}(\varsigma) = I - \left(1 - \frac{\varsigma_0 - \varsigma}{1 - \bar{\varsigma}_0\varsigma}\frac{|\varsigma_0|}{\varsigma_0}\right)p,$$

$$p = \begin{bmatrix}\xi^* \\ \eta^*\end{bmatrix}[\xi, \eta]j = \begin{bmatrix}-\xi^*\xi & \xi^*\eta \\ -\eta^*\xi & \eta^*\eta\end{bmatrix},$$

$$-\xi\xi^* + \eta\eta^* = I,$$

is an arbitrary binomial j-contractive elementary factor of full rank.

We represent the matrices ξ and η in polar form, $\xi = |\xi|u$ and $\eta = |\eta|v$, where $|\varsigma| = \sqrt{\xi\xi^*}$, $|\eta| = \sqrt{\eta\eta^*}$, and u and v are unitary matrices, and we consider the new matrix-valued function

$$\hat{f}^{-1}(\varsigma) = I - \left(1 - \frac{\varsigma_0 - \varsigma}{1 - \bar{\varsigma}_0\varsigma}\frac{|\varsigma_0|}{\varsigma_0}\right)\hat{p},$$

$$\hat{p} = \begin{bmatrix}|\xi| \\ |\eta|\end{bmatrix}[|\xi|, |\eta|]j.$$

The main condition for $f^{-1}(\varsigma)$ now implies the equality

$$-|\xi||\xi| + |\eta||\eta| = I$$

and, consequently, $\hat{f}^{-1}(\varsigma)$ is an elementary j-contractive factor for which the additional condition (9) coincides with the main condition. Lemma 1 is applicable to $\hat{f}^{-1}(\varsigma)$, and thus

$$\hat{\rho}_d = \hat{r}^{-1} = [I + (1 - |\varsigma_0|^2)|\xi|^2]^{-1},$$

$$\hat{\rho}_g = \hat{s}\hat{r}^{-1}\hat{s}^* - \hat{t} = |\varsigma_0|^2\hat{r}^{-1}.$$

The auxiliary elementary matrix-valued function $\hat{f}^{-1}(\varsigma)$, which we say is *central with respect to* $f^{-1}(\varsigma)$, plays an essential role in the subsequent constructions.

To prove the next theorems we need an elementary lemma.

LEMMA 2. *If the block A in the matrix $\begin{bmatrix}A & B \\ C & D\end{bmatrix}$ is nonsingular, then:*
1) *It admits a "triangular" representation*

$$\begin{bmatrix}A & B \\ C & D\end{bmatrix} = \begin{bmatrix}I & 0 \\ CA^{-1} & I\end{bmatrix}\begin{bmatrix}A & 0 \\ 0 & D - CA^{-1}B\end{bmatrix}\begin{bmatrix}I & A^{-1}B \\ 0 & I\end{bmatrix}.$$

2) *This representation is unique.*

The first assertion can be checked directly.

If we have the equality

$$\begin{bmatrix}I & 0 \\ X & I\end{bmatrix}\begin{bmatrix}A & 0 \\ 0 & \varepsilon\end{bmatrix}\begin{bmatrix}I & Y \\ 0 & I\end{bmatrix} = \begin{bmatrix}I & 0 \\ X_1 & I\end{bmatrix}\begin{bmatrix}A_1 & 0 \\ 0 & \varepsilon_1\end{bmatrix}\begin{bmatrix}I & Y_1 \\ 0 & I\end{bmatrix},$$

then it can be written in the form

$$\begin{bmatrix}I & 0 \\ X - X_1 & I\end{bmatrix}\begin{bmatrix}A & 0 \\ 0 & \varepsilon\end{bmatrix}\begin{bmatrix}I & Y - Y_1 \\ 0 & I\end{bmatrix} = \begin{bmatrix}A_1 & 0 \\ 0 & \varepsilon_1\end{bmatrix},$$

or

$$\begin{bmatrix}A & A(Y - Y_1) \\ (X - X_1)A & (X - X_1)A(Y - Y_1) + \varepsilon\end{bmatrix} = \begin{bmatrix}A_1 & 0 \\ 0 & \varepsilon_1\end{bmatrix}.$$

Since A is nonsingular, the latter implies the relations $A_1 = A$, $X_1 = X$, $Y_1 = Y$, and $\varepsilon_1 = \varepsilon$, which prove the second assertion of the lemma.

THEOREM 2. *In the j-metric the right and left radii of the Weyl disk for an arbitrary j-contractive binomial factor $f^{-1}(\varsigma)$ of full rank are equal to*

$$\rho_d = r^{-1} = u^* \hat{r}^{-1} u,$$

$$\rho_g = s r^{-1} s^* - t = |\varsigma_0|^2 v^* \hat{r}^{-1} v,$$

respectively, where u and v are the unitary matrices in the polar representation $\xi = |\xi| u$, $\eta = |\eta| v$, and $\hat{r}^{-1} = [I + (1 - |\varsigma_0|^2)|\xi|^2]^{-1}$ is the right radius of the factor $\hat{f}^{-1}(\varsigma)$ which is central with respect to $f^{-1}(\varsigma)$.

PROOF. Indeed, since

$$f^{-1}(\varsigma) = \begin{bmatrix} u^* & 0 \\ 0 & v^* \end{bmatrix} \hat{f}^{-1}(\varsigma) \begin{bmatrix} u & 0 \\ 0 & v \end{bmatrix},$$

the Weyl matrix for $f^{-1}(\varsigma)$ is equal to

$$w = f^{-1}(0) j f^{*-1}(0) = \begin{bmatrix} u^* & 0 \\ 0 & v^* \end{bmatrix} \hat{w} \begin{bmatrix} u & 0 \\ 0 & v \end{bmatrix}$$

$$= \begin{bmatrix} u^* & 0 \\ 0 & v^* \end{bmatrix} \begin{bmatrix} I & 0 \\ -\hat{s}\hat{r}^{-1} & I \end{bmatrix} \begin{bmatrix} -\hat{r} & 0 \\ 0 & \hat{s}\hat{r}^{-1}\hat{s}^* - \hat{t} \end{bmatrix} \begin{bmatrix} I & -\hat{r}^{-1}\hat{s}^* \\ 0 & I \end{bmatrix} \begin{bmatrix} u & 0 \\ 0 & v \end{bmatrix}$$

$$= \begin{bmatrix} u^* & 0 \\ 0 & v^* \end{bmatrix} \begin{bmatrix} I & 0 \\ -\hat{s}\hat{r}^{-1} & I \end{bmatrix} \begin{bmatrix} -\hat{r} & 0 \\ 0 & |\varsigma_0|^2 \hat{r}^{-1} \end{bmatrix} \begin{bmatrix} I & -\hat{r}^{-1}\hat{s}^* \\ 0 & I \end{bmatrix} \begin{bmatrix} u & 0 \\ 0 & v \end{bmatrix}.$$

If in the last expression we transpose the first and second factors and the fourth and fifth factors and then use the obvious identity

$$\begin{bmatrix} I & -\hat{r}^{-1}\hat{s}^* \\ 0 & I \end{bmatrix} \begin{bmatrix} u & 0 \\ 0 & v \end{bmatrix} = \begin{bmatrix} u & 0 \\ 0 & v \end{bmatrix} \begin{bmatrix} I & -u^* \hat{r}^{-1}\hat{s}^* v \\ 0 & I \end{bmatrix},$$

we get that

$$w = \begin{bmatrix} I & 0 \\ -v^* \hat{s}\hat{r}^{-1} u & I \end{bmatrix} \begin{bmatrix} u^* & 0 \\ 0 & v^* \end{bmatrix} \begin{bmatrix} -\hat{r} & 0 \\ 0 & |\varsigma_0|^2 \hat{r}^{-1} \end{bmatrix} \begin{bmatrix} u & 0 \\ 0 & v \end{bmatrix} \begin{bmatrix} I & -u^* \hat{r}^{-1}\hat{s}^* v \\ 0 & I \end{bmatrix}$$

$$= \begin{bmatrix} I & 0 \\ -v^* \hat{s}\hat{r}^{-1} u & I \end{bmatrix} \begin{bmatrix} -u^* \hat{r} u & 0 \\ 0 & |\varsigma_0|^2 v^* \hat{r}^{-1} v \end{bmatrix} \begin{bmatrix} I & -u^* \hat{r}^{-1}\hat{s}^* v \\ 0 & I \end{bmatrix}.$$

By assertion 2) of Lemma 2, $u^* \hat{r}^{-1} u$ and $|\varsigma_0|^2 v^* \hat{r}^{-1} v$ are the right and left radii of the Weyl disk for $f^{-1}(\varsigma)$. The theorem is proved.

Note that $0 < \hat{r}^{-1} \le I$, since $\hat{r}^{-1} = [I + (1 - |\varsigma_0|^2)\xi^* \xi]^{-1}$.

The converse assertion will play an essential role for us.

THEOREM 3. *For any nonsingular positive Hermitian matrix $0 < r^{-1} \le I$ and any unitary matrices u and v there exists a family of elementary binomial j-contractive factors $f^{-1}(\varsigma)$ of full rank such that the matrices $u^* r^{-1} u$ and $|\varsigma_0|^2 v^* r^{-1} v$ are the right and left radii of the Weyl disk, respectively.*

PROOF. First of all, from the radii $\tilde{\rho}_d = r^{-1}$ and $\tilde{\rho}_g = |\varsigma_0|^2 r^{-1}$ at the point $\varsigma = 0$ we construct a family of factors $\tilde{f}^{-1}(\varsigma)$ satisfying the additional condition that $-\xi^* \xi + \eta^* \eta = I$, as done in the second part of Lemma 1:

$$\tilde{f}^{-1}(\varsigma) = I - \left(1 - \frac{\varsigma_0 - \varsigma}{1 - \bar{\varsigma}_0 \varsigma} \frac{|\varsigma_0|}{\varsigma_0}\right) p;$$

$$p = \begin{bmatrix} \dfrac{r-1}{1-|\varsigma_0|^2} & \dfrac{\sqrt{r-I}\,z\sqrt{r-|\varsigma_0|^2 I}}{1-|\varsigma_0|^2} \\[4mm] \dfrac{\sqrt{r-|\varsigma_0|^2 I}\,z\sqrt{r-I}}{1-|\varsigma_0|^2} & \dfrac{r-|\varsigma_0|^2 I}{1-|\varsigma_0|^2} \end{bmatrix} j,$$

where the unitary matrix z commuting with r is a parameter.

Finally, as in Theorem 2, we show that the radii of the Weyl disk for the factors

$$f^{-1}(\varsigma) = \begin{bmatrix} u^* & 0 \\ 0 & v^* \end{bmatrix} \tilde{f}^{-1}(\varsigma) \begin{bmatrix} u & 0 \\ 0 & v \end{bmatrix}$$

are precisely equal to $\rho_d = u^* r^{-1} u$ and $\rho_g = |\varsigma_0|^2 v^* r^{-1} v$, which is what was required.

§**2.** Let us consider the computation of the radii of the Weyl disk for the finite product

$$\mathfrak{A}_k^{-1}(\varsigma) = f_1^{-1}(\varsigma) f_2^{-2}(\varsigma) \cdots f_k^{-1}(\varsigma)$$

of elementary binomial J-contractive factors of full rank. These radii are computed at the point $\varsigma = 0$ under the assumption that $\varsigma = 0$ is not a node of interpolation. The process will be implemented step by step by looking at the product of two, three, etc., factors.

Suppose first that

$$\mathfrak{A}_1^{-1}(\varsigma) = f_1^{-1}(\varsigma), \qquad f_1^{-1}(\varsigma) = I - (1 - |\varsigma_1|^2) P_1;$$

$$P_1 = \begin{bmatrix} x_1^* \\ y_1^* \end{bmatrix} [x_1, y_1] J, \qquad [x_1, y_1] J \begin{bmatrix} x_1^* \\ y_1^* \end{bmatrix} = I.$$

We form the Weyl matrix

$$\mathfrak{A}_1^{-1}(0) J \mathfrak{A}^{*-1}(0) = W_1 = \begin{bmatrix} -R_1 & S_1^* \\ S_1 & T_1 \end{bmatrix}. \tag{19}$$

The computations below are a preliminary step for the consideration of a product of two factors:

$$W_1 = \begin{bmatrix} -R_1 & S_1^* \\ S_1 & -T_1 \end{bmatrix} = \begin{bmatrix} I & 0 \\ -S_1 R_1^{-1} & I \end{bmatrix} \begin{bmatrix} -R_1 & 0 \\ 0 & S_1 R_1^{-1} S_1^* - T_1 \end{bmatrix} \begin{bmatrix} I & -R_1^1 S_1^* \\ 0 & I \end{bmatrix}$$

$$= \begin{bmatrix} I & 0 \\ -S_1 R_1^{-1} & I \end{bmatrix} \begin{bmatrix} -R_1 & 0 \\ 0 & |\varsigma_1|^2 R_1^{-1} \end{bmatrix} \begin{bmatrix} I & -R_1^{-1} S_1^* \\ 0 & I \end{bmatrix}$$

$$= \begin{bmatrix} I & 0 \\ -S_1 R_1^{-1} & I \end{bmatrix} \begin{bmatrix} R^{1/2} & 0 \\ 0 & |\varsigma_1| R_1^{-1/2} \end{bmatrix} \begin{bmatrix} -I & 0 \\ 0 & I \end{bmatrix} \begin{bmatrix} R^{1/2} & 0 \\ 0 & |\varsigma_1| R^{-1/2} \end{bmatrix} \begin{bmatrix} I & -R_1^{-1} S_1^* \\ 0 & I \end{bmatrix},$$

but since

$$\begin{bmatrix} -I & 0 \\ 0 & I \end{bmatrix} = \begin{bmatrix} \frac{1}{\sqrt{2}} I & -\frac{1}{\sqrt{2}} I \\ \frac{1}{\sqrt{2}} I & \frac{1}{\sqrt{2}} I \end{bmatrix} \begin{bmatrix} 0 & I \\ I & 0 \end{bmatrix} \begin{bmatrix} \frac{1}{\sqrt{2}} I & \frac{1}{\sqrt{2}} I \\ -\frac{1}{\sqrt{2}} I & \frac{1}{\sqrt{2}} I \end{bmatrix},$$

it follows that

$$W_1 = \begin{bmatrix} I & 0 \\ -S_1 R_1^{-1} & I \end{bmatrix} \begin{bmatrix} R_1^{1/2} & 0 \\ 0 & |\varsigma_1| R_1^{-1/2} \end{bmatrix} \begin{bmatrix} \frac{1}{\sqrt{2}} I & -\frac{1}{\sqrt{2}} I \\ \frac{1}{\sqrt{2}} I & \frac{1}{\sqrt{2}} I \end{bmatrix} J$$

$$\times \begin{bmatrix} \frac{1}{\sqrt{2}} I & \frac{1}{\sqrt{2}} I \\ -\frac{1}{\sqrt{2}} I & \frac{1}{\sqrt{2}} I \end{bmatrix} \begin{bmatrix} R^{1/2} & 0 \\ 0 & |\varsigma_1| R_1^{-1/2} \end{bmatrix} \begin{bmatrix} I & -R_1^{-1} S_1^* \\ 0 & I \end{bmatrix}. \tag{20}$$

Comparing (19) and (20) now, we arrive at the following expression for $\mathfrak{A}_1^{-1}(0)$:

$$\mathfrak{A}_1^{-1}(0) = \begin{bmatrix} I & 0 \\ -S_1 R_1^{-1} & I \end{bmatrix} \begin{bmatrix} R_1^{1/2} & 0 \\ 0 & |\varsigma_1| R_1^{-1/2} \end{bmatrix} \begin{bmatrix} \frac{1}{\sqrt{2}} I & -\frac{1}{\sqrt{2}} I \\ \frac{1}{\sqrt{2}} I & \frac{1}{\sqrt{2}} I \end{bmatrix} \mathfrak{Y}_1, \qquad (21)$$

where $\mathfrak{Y}_1$ is a J-unitary matrix, $\mathfrak{Y}_1 J \mathfrak{Y}_1^* = J$, which can be computed by the formula

$$\mathfrak{Y}_1 = \begin{bmatrix} \frac{1}{\sqrt{2}} I & \frac{1}{\sqrt{2}} I \\ -\frac{1}{\sqrt{2}} I & \frac{1}{\sqrt{2}} I \end{bmatrix} \begin{bmatrix} R_1^{-1/2} & 0 \\ 0 & \frac{1}{|\varsigma_1|} R_1^{1/2} \end{bmatrix} \begin{bmatrix} I & 0 \\ S_1 R_1^{-1} & I \end{bmatrix} \mathfrak{A}_1^{-1}(0). \qquad (22)$$

Suppose now that

$$\mathfrak{A}_2^{-1}(\varsigma) = \mathfrak{A}_1^{-1}(\varsigma) f_2^{-1}(\varsigma), \qquad f_2^{-1}(0) = I - (1 - |\varsigma_2|^2) P_2,$$

$$P_2 = \begin{bmatrix} x_2^* \\ y_2^* \end{bmatrix} [x_2, y_2] J, \qquad [x_2, y_2] J \begin{bmatrix} x_2^* \\ y_2^* \end{bmatrix} = I.$$

We use (21) for expressing the product,

$$\mathfrak{A}_2^{-1}(0) = \mathfrak{A}_1^{-1}(0) f_2^{-1}(0)$$

$$= \begin{bmatrix} I & 0 \\ -S_1 R_1^{-1} & I \end{bmatrix} \begin{bmatrix} R_1^{1/2} & 0 \\ 0 & |\varsigma_1| R_1^{-1/2} \end{bmatrix} \begin{bmatrix} \frac{1}{\sqrt{2}} I & -\frac{1}{\sqrt{2}} I \\ \frac{1}{\sqrt{2}} I & \frac{1}{\sqrt{2}} I \end{bmatrix} \mathfrak{Y}_1 f_2^{-1}(0)$$

and consider the Weyl matrix

$$W_2 = \mathfrak{A}_2^{-1}(0) J \mathfrak{A}_2^{*-1}(0)$$

$$= \begin{bmatrix} I & 0 \\ -S_1 R_1^{-1} & I \end{bmatrix} \begin{bmatrix} R_1^{1/2} & 0 \\ 0 & |\varsigma_1| R_1^{-1/2} \end{bmatrix} \begin{bmatrix} \frac{1}{\sqrt{2}} I & -\frac{1}{\sqrt{2}} I \\ \frac{1}{\sqrt{2}} I & \frac{1}{\sqrt{2}} I \end{bmatrix} \mathfrak{Y}_1 f_2^{-1}(0) J f_2^{*-1}(0) \mathfrak{Y}_1^*$$

$$\times \begin{bmatrix} \frac{1}{\sqrt{2}} I & \frac{1}{\sqrt{2}} I \\ -\frac{1}{\sqrt{2}} I & \frac{1}{\sqrt{2}} I \end{bmatrix} \begin{bmatrix} R_1^{1/2} & 0 \\ 0 & |\varsigma_1| R_1^{-1/2} \end{bmatrix} \begin{bmatrix} I & -R_1^{-1} S_1^* \\ 0 & I \end{bmatrix}.$$

The last expression can be rewritten in a convenient form if J is replaced by

$$\mathfrak{Y}_1^{-1} \begin{bmatrix} \frac{1}{\sqrt{2}} I & \frac{1}{\sqrt{2}} I \\ -\frac{1}{\sqrt{2}} I & \frac{1}{\sqrt{2}} I \end{bmatrix} \begin{bmatrix} \frac{1}{\sqrt{2}} I & -\frac{1}{\sqrt{2}} I \\ \frac{1}{\sqrt{2}} I & \frac{1}{\sqrt{2}} I \end{bmatrix} J \begin{bmatrix} \frac{1}{\sqrt{2}} I & \frac{1}{\sqrt{2}} I \\ -\frac{1}{\sqrt{2}} I & \frac{1}{\sqrt{2}} I \end{bmatrix} \begin{bmatrix} \frac{1}{\sqrt{2}} I & -\frac{1}{\sqrt{2}} I \\ \frac{1}{\sqrt{2}} I & \frac{1}{\sqrt{2}} I \end{bmatrix} \mathfrak{Y}_1^{*-1}.$$

Then

$$W_2 = \begin{bmatrix} I & 0 \\ -S_1 R_1^{-1} & I \end{bmatrix} \begin{bmatrix} R_1^{1/2} & 0 \\ 0 & |\varsigma_1| R_1^{-1/2} \end{bmatrix} \begin{bmatrix} \frac{1}{\sqrt{2}} I & -\frac{1}{\sqrt{2}} I \\ \frac{1}{\sqrt{2}} I & \frac{1}{\sqrt{2}} I \end{bmatrix} \mathfrak{Y}_1 f_2^{-1}(0) \mathfrak{Y}_1^{-1} \begin{bmatrix} \frac{1}{\sqrt{2}} I & \frac{1}{\sqrt{2}} I \\ -\frac{1}{\sqrt{2}} I & \frac{1}{\sqrt{2}} I \end{bmatrix}$$

$$\times \begin{bmatrix} \frac{1}{\sqrt{2}} I & -\frac{1}{\sqrt{2}} I \\ \frac{1}{\sqrt{2}} I & \frac{1}{\sqrt{2}} I \end{bmatrix} J \begin{bmatrix} \frac{1}{\sqrt{2}} I & \frac{1}{\sqrt{2}} I \\ -\frac{1}{\sqrt{2}} I & \frac{1}{\sqrt{2}} I \end{bmatrix}$$

$$\times \begin{bmatrix} \frac{1}{\sqrt{2}} I & -\frac{1}{\sqrt{2}} I \\ \frac{1}{\sqrt{2}} I & \frac{1}{\sqrt{2}} I \end{bmatrix} \mathfrak{Y}_1^{*-1} f_2^{*-1}(0) \mathfrak{Y}_1^* \begin{bmatrix} \frac{1}{\sqrt{2}} I & \frac{1}{\sqrt{2}} I \\ -\frac{1}{\sqrt{2}} I & \frac{1}{\sqrt{2}} I \end{bmatrix} \begin{bmatrix} R_1^{1/2} & 0 \\ 0 & |\varsigma_1| R_1^{-1/2} \end{bmatrix} \begin{bmatrix} I & -R_1^{-1} S_1^* \\ 0 & I \end{bmatrix}.$$

$$(23)$$

We introduce the new matrix-valued function

$$\beta_2^{-1}(\varsigma) = \begin{bmatrix} \frac{1}{\sqrt{2}} I & -\frac{1}{\sqrt{2}} I \\ \frac{1}{\sqrt{2}} I & \frac{1}{\sqrt{2}} I \end{bmatrix} \mathfrak{Y}_1 f_2^{-1}(\varsigma) \mathfrak{Y}_1^{-1} \begin{bmatrix} \frac{1}{\sqrt{2}} I & \frac{1}{\sqrt{2}} I \\ -\frac{1}{\sqrt{2}} I & \frac{1}{\sqrt{2}} I \end{bmatrix}$$

$$= I - \left(1 - \frac{\varsigma_2 \overset{\cdot}{-} \varsigma}{1 - \bar{\varsigma}_2 \varsigma} \frac{|\varsigma_2|}{\varsigma_2} \right) p_2; \qquad (24)$$

$$p_2 = \begin{bmatrix} \frac{1}{\sqrt{2}}I & -\frac{1}{\sqrt{2}}I \\ \frac{1}{\sqrt{2}}I & \frac{1}{\sqrt{2}}I \end{bmatrix} \mathfrak{Y}_1 p_2 \mathfrak{Y}_1^{-1} \begin{bmatrix} \frac{1}{\sqrt{2}}I & \frac{1}{\sqrt{2}}I \\ -\frac{1}{\sqrt{2}}I & \frac{1}{\sqrt{2}}I \end{bmatrix} = \begin{bmatrix} \xi_2^* \\ \eta_2^* \end{bmatrix} [\xi_2, \eta_2]j,$$

$$[\xi_2, \eta_2]j \begin{bmatrix} \xi_2^* \\ \eta_2^* \end{bmatrix} = I.$$

Obviously, $\beta_2^{-1}(\varsigma)$ is an elementary j-contractive factor whose Weyl matrix, by Theorem 2, has the form

$$\begin{aligned} w_2 &= \beta_2^{-1}(0)j\beta_2^{*-1}(0) \\ &= \begin{bmatrix} I & 0 \\ -v_2^*\hat{s}_2\hat{r}_2^{-1}u_2 & I \end{bmatrix} \begin{bmatrix} -u_2^*\hat{r}_2u_2 & 0 \\ 0 & |\varsigma_2|^2 v_2^*\hat{r}_2^{-1}v_2 \end{bmatrix} \begin{bmatrix} I & -u_2^*\hat{r}_2^{-1}\hat{s}_2^*v_2 \\ 0 & I \end{bmatrix}. \end{aligned}$$

Here $\hat{r}_2 = I + (1 - |\varsigma_2|^2)|\xi_2|^2$, $\xi_2 = |\xi_2|u_2$, and $\eta_2 = |\eta_2|v_2$. Considering (24), we now rewrite equality (23) for W_2 in the form

$$\begin{aligned} W_2 &= \begin{bmatrix} I & 0 \\ -S_1R_1^{-1} & I \end{bmatrix} \begin{bmatrix} R_1^{1/2} & 0 \\ 0 & |\varsigma_1|R_1^{-1/2} \end{bmatrix} \beta_2^{-1}(0)j\beta_2^{*-1}(0) \\ &\qquad\qquad \times \begin{bmatrix} R_1^{1/2} & 0 \\ 0 & |\varsigma_1|R_1^{-1/2} \end{bmatrix} \begin{bmatrix} I & -R_1^{-1}S_1^* \\ 0 & I \end{bmatrix}, \end{aligned}$$

or

$$\begin{aligned} W_2 &= \begin{bmatrix} I & 0 \\ -S_1R_1^{-1} & I \end{bmatrix} \begin{bmatrix} R_1^{1/2} & 0 \\ 0 & |\varsigma_1|R_1^{-1/2} \end{bmatrix} w_2 \begin{bmatrix} R_1^{1/2} & 0 \\ 0 & |\varsigma_1|R_1^{-1/2} \end{bmatrix} \begin{bmatrix} I & -R_1^{-1}S_1^* \\ 0 & I \end{bmatrix} \\ &= \begin{bmatrix} I & 0 \\ -S_1R_1^{-1} & I \end{bmatrix} \begin{bmatrix} R_1^{1/2} & 0 \\ 0 & |\varsigma_1|R_1^{-1/2} \end{bmatrix} \begin{bmatrix} I & 0 \\ -v_2^*\hat{s}_2\hat{r}_2^{-1}u_2 & I \end{bmatrix} \begin{bmatrix} -u_2^*\hat{r}_2u_2 & 0 \\ 0 & |\varsigma_2|^2 v_2^*\hat{r}_2^{-1}v_2 \end{bmatrix} \\ &\qquad \times \begin{bmatrix} I & -u_2^*\hat{r}_2^{-1}\hat{s}_2^*v_2 \\ 0 & I \end{bmatrix} \begin{bmatrix} R_1^{1/2} & 0 \\ 0 & |\varsigma_1|R_1^{-1/2} \end{bmatrix} \begin{bmatrix} I & -R_1^{-1}S_1^* \\ 0 & I \end{bmatrix}. \end{aligned}$$

Finally, transposing the second and third factors and the fifth and sixth factors of the last product, we get that

$$\begin{aligned} W_2 &= \begin{bmatrix} I & 0 \\ -S_1R_1^{-1} & I \end{bmatrix} \begin{bmatrix} I & 0 \\ R_1^{-1/2}v_2^*\hat{s}_2\hat{r}_2^{-1}u_2R_1^{-1/2}|\varsigma_1| & I \end{bmatrix} \\ &\quad \times \begin{bmatrix} R_1^{1/2} & 0 \\ 0 & |\varsigma_1|R_1^{-1/2} \end{bmatrix} \begin{bmatrix} -u_2^*\hat{r}_2u_2 & 0 \\ 0 & |\varsigma_2|^2 v_2^*\hat{r}_2^{-1}v_2 \end{bmatrix} \begin{bmatrix} R_1^{1/2} & 0 \\ 0 & |\varsigma_1|R_1^{-1/2} \end{bmatrix} \\ &\quad \times \begin{bmatrix} I & -|\varsigma_1|R_1^{-1/2}u_2^*\hat{r}_2^{-1}\hat{s}_2^*v_2R_1^{-1/2} \\ 0 & I \end{bmatrix} \begin{bmatrix} I & -R_1^{-1}S_1^* \\ 0 & I \end{bmatrix}, \end{aligned}$$

which gives us, finally, that

$$W_2 = \begin{bmatrix} I & 0 \\ -S_2R_2^{-1} & I \end{bmatrix} \begin{bmatrix} -R_1^{1/2}u_2^*\hat{r}_2^{-1}u_2R_1^{1/2} & 0 \\ 0 & |\varsigma_1|^2|\varsigma_2|^2 R_1^{-1/2}v_2^*\hat{r}_2^{-1}v_2R_1^{-1/2} \end{bmatrix} \begin{bmatrix} I & -R_2^{-1}S_2^* \\ 0 & I \end{bmatrix}.$$

On the other hand, since

$$W_2 = \begin{bmatrix} -R_2 & S_2^* \\ S_2 & -T_2 \end{bmatrix} = \begin{bmatrix} I & 0 \\ -S_2R_2^{-1} & I \end{bmatrix} \begin{bmatrix} -\rho_{d2}^{-1} & 0 \\ 0 & \rho_{g2} \end{bmatrix} \begin{bmatrix} I & -R_2^{-1}S_2^* \\ 0 & I \end{bmatrix},$$

where $\rho_{d2} = R_2^{-1}$ and $\rho_{g2} = S_2 R_2^{-1} S_2^* - T_2$, it follows from the assertion in Lemma 2 on the uniqueness of the triangular representation that

$$\rho_{d2} = R_1^{-1/2} u_2^* \hat{r}_2^{-1} u_2 R_1^{-1/2}, \qquad \rho_{g2} = |\varsigma_1|^2 |\varsigma_2|^2 R_1^{-1/2} v_2^* \hat{r}_2^{-1} v_2 R_1^{-1/2}.$$

The process described here can be continued indefinitely. To compute the radii of the Weyl disk for a product of three factors,

$$\mathfrak{A}_3^{-1}(0) = f_1^{-1}(0) f_2^{-1}(0) f_3^{-1}(0),$$

we transform $\mathfrak{A}_2^{-1}(0)$ to a convenient form with the help of the Weyl matrix

$$
\begin{aligned}
W_2 = \mathfrak{A}_2^{-1}(0) J \mathfrak{A}_2^{*-1}(0) &= \begin{bmatrix} I & 0 \\ -S_2 R_2^{-1} & I \end{bmatrix} \begin{bmatrix} -\rho_{d2}^{-1} & 0 \\ 0 & \rho_{g2}^{-1} \end{bmatrix} \begin{bmatrix} I & -R_2^{-1} S_2^* \\ 0 & I \end{bmatrix} \\
&= \begin{bmatrix} I & 0 \\ -S_2 R_2^{-1} & I \end{bmatrix} \begin{bmatrix} R_1^{1/2} u_2^* \hat{r}_2^{1/2} & 0 \\ 0 & |\varsigma_1||\varsigma_2| R_1^{-1/2} v_2^* \hat{r}_2^{-1/2} \end{bmatrix} \begin{bmatrix} -I & 0 \\ 0 & I \end{bmatrix} \\
&\qquad\qquad \times \begin{bmatrix} \hat{r}_2^{1/2} u_2 R_1^{1/2} & 0 \\ 0 & |\varsigma_1||\varsigma_2| \hat{r}_2^{-1/2} v_2 R_1^{-1/2} \end{bmatrix} \begin{bmatrix} I & -R_2^{-1} S_2^* \\ 0 & I \end{bmatrix} \\
&= \begin{bmatrix} I & 0 \\ -S_2 R_2^{-1} & I \end{bmatrix} \begin{bmatrix} R_1^{1/2} u_2^* \hat{r}_2^{1/2} & 0 \\ 0 & |\varsigma_1||\varsigma_2| R_1^{-1/2} v_2^* \hat{r}_2^{-1/2} \end{bmatrix} \begin{bmatrix} \frac{1}{\sqrt{2}} I & -\frac{1}{\sqrt{2}} I \\ \frac{1}{\sqrt{2}} I & \frac{1}{\sqrt{2}} I \end{bmatrix} J \\
&\qquad \times \begin{bmatrix} \frac{1}{\sqrt{2}} I & \frac{1}{\sqrt{2}} I \\ -\frac{1}{\sqrt{2}} I & \frac{1}{\sqrt{2}} I \end{bmatrix} \begin{bmatrix} \hat{r}_2^{1/2} u_2 R_1^{1/2} & 0 \\ 0 & |\varsigma_1||\varsigma_2| \hat{r}_2^{1/2} v_2 R_1^{-1/2} \end{bmatrix} \begin{bmatrix} I & -R_2^{-1} S_2^* \\ 0 & I \end{bmatrix},
\end{aligned}
$$

from which it is determined that

$$
\begin{aligned}
\mathfrak{A}_2^{-1}(0) = \begin{bmatrix} I & 0 \\ -S_2 R_2^{-1} & I \end{bmatrix} &\begin{bmatrix} R_1^{1/2} u_2^* \hat{r}_2^{1/2} & 0 \\ 0 & |\varsigma_1||\varsigma_2| R_1^{-1/2} v_2^* \hat{r}_2^{-1/2} \end{bmatrix} \\
&\times \begin{bmatrix} \frac{1}{\sqrt{2}} I & -\frac{1}{\sqrt{2}} I \\ \frac{1}{\sqrt{2}} I & \frac{1}{\sqrt{2}} I \end{bmatrix} \mathfrak{Y}_2.
\end{aligned} \tag{25}
$$

Here $\mathfrak{Y}_2$ is a J-unitary matrix, $\mathfrak{Y}_2 J \mathfrak{Y}_2^* = J$, which is computed by the formula

$$
\begin{aligned}
\mathfrak{Y}_2 = \begin{bmatrix} \frac{1}{\sqrt{2}} I & \frac{1}{\sqrt{2}} I \\ -\frac{1}{\sqrt{2}} I & \frac{1}{\sqrt{2}} I \end{bmatrix} &\begin{bmatrix} \hat{r}_2^{-1/2} u_2 R_1^{-1/2} & 0 \\ 0 & \dfrac{1}{|\varsigma_1||\varsigma_2|} \hat{r}_2^{1/2} v_2 R_1^{1/2} \end{bmatrix} \\
&\times \begin{bmatrix} I & 0 \\ S_2 R_2^{-1} & I \end{bmatrix} \mathfrak{A}_2^{-1}(0).
\end{aligned} \tag{26}
$$

Knowing $\mathfrak{Y}_2$, we pass from $f^{-1}(\varsigma)$ to the j-contractive factor

$$
\begin{aligned}
\beta_3^{-1}(\varsigma) &= \begin{bmatrix} \frac{1}{\sqrt{2}} I & -\frac{1}{\sqrt{2}} I \\ \frac{1}{\sqrt{2}} I & \frac{1}{\sqrt{2}} I \end{bmatrix} \mathfrak{Y}_2 f_3^{-1}(\varsigma) \mathfrak{Y}_2^{-1} \begin{bmatrix} \frac{1}{\sqrt{2}} I & \frac{1}{\sqrt{2}} I \\ -\frac{1}{\sqrt{2}} I & \frac{1}{\sqrt{2}} I \end{bmatrix} \\
&= I - \left(1 - \frac{\varsigma_3 - \varsigma}{1 - \varsigma_3 \varsigma} \frac{|\varsigma_3|}{\varsigma_3} \right) p_3,
\end{aligned}
$$

where

$$p_3 = \begin{bmatrix} \frac{1}{\sqrt{2}}I & -\frac{1}{\sqrt{2}}I \\ \frac{1}{\sqrt{2}}I & \frac{1}{\sqrt{2}}I \end{bmatrix} \mathfrak{Y}_2 p_3 \mathfrak{Y}_2^{-1} \begin{bmatrix} \frac{1}{\sqrt{2}}I & \frac{1}{\sqrt{2}}I \\ -\frac{1}{\sqrt{2}}I & \frac{1}{\sqrt{2}}I \end{bmatrix} = \begin{bmatrix} \xi_3^* \\ \eta_3^* \end{bmatrix} [\xi_3, \eta_3] j,$$

$$[\xi_3, \eta_3] j \begin{bmatrix} \xi_3^* \\ \eta_3^* \end{bmatrix} = I, \qquad p_3^2 = p_3, \qquad p_3 j \geq 0.$$

Repeating the preceding computations word for word, we get for the Weyl matrix W_3 the equality

$$W_3 = \begin{bmatrix} I & 0 \\ -S_3 R_3^{-1} & I \end{bmatrix}$$
$$\times \begin{bmatrix} -R_1^{1/2} u_2^* \hat{r}_2^{1/2} u_3^* \hat{r}_3 u_3 \hat{r}_2^{1/2} u_2 R_2^{1/2} & 0 \\ 0 & |\varsigma_1|^2 |\varsigma_2|^2 |\varsigma_3|^2 R_1^{-1/2} v_2^* \hat{r}_2^{-1/2} v_3^* \hat{r}_3^{-1} v_3 \hat{r}_2^{-1/2} v_2 R_1^{-1/2} \end{bmatrix}$$
$$\times \begin{bmatrix} I & -R_3^{-1} S_3^* \\ 0 & I \end{bmatrix},$$

which gives us the relations

$$\rho_{d3} = R_1^{-1/2} u_2^* \hat{r}_2^{-1/2} u_3^* \hat{r}_3^{-1} u_3 \hat{r}_2^{-1/2} u_2 R_2^{-1/2},$$
$$\rho_{g3} = |\varsigma_1|^2 |\varsigma_2|^2 |\varsigma_3|^2 R_1^{-1/2} v_2^* r_2^{-1/2} v_3^* \hat{r}_3^{-1} v_3 \hat{r}_2^{-1/2} v_2 R_1^{-1/2}$$

for the radii.

Continuing, we get the following expressions for the radii of the Weyl disk corresponding to the product of k binomial J-contractive factors of full rank:

$$\mathfrak{A}_k^{-1}(\varsigma) = f_1^{-1}(\varsigma) f_2^{-1}(\varsigma) \cdots f_k^{-1}(\varsigma);$$
$$\rho_{dk} = R_1^{-1/2} u_2^* \hat{r}_2^{1/2} u_3^* \hat{r}_3^{-1/2} \cdots u_k^* \hat{r}_k^{-1} u_k \cdots \hat{r}_3^{-1/2} u_3 \hat{r}_2^{-1/2} u_2 R_1^{-1/2};$$
$$\rho_{gk} = |\varsigma_1|^2 |\varsigma_2|^2 \cdots |\varsigma_k|^2 R_1^{-1/2} v_2^* \hat{r}_2^{-1/2} v_3^* \hat{r}_3^{-1/2}$$
$$\cdots v_k^* \hat{r}_k^{-1} v_k \cdots \hat{r}_3^{-1/2} v_3 \hat{r}_2^{-1/2} v_2 R_1^{-1/2}.$$

This proves the following basic theorem.

THEOREM 4. *Let*

$$\mathfrak{A}_k^{-1}(\varsigma) = f_1^{-1}(\varsigma) f_2^{-1}(\varsigma) \cdots f_k^{-1}(\varsigma)$$
$$= \prod_{\nu=1}^{k} \stackrel{\curvearrowright}{} f_\nu^{-1}(\varsigma) = \prod_{\nu=1}^{k} \stackrel{\curvearrowright}{} \left[I - \left(1 - \frac{\varsigma_\nu - \varsigma}{1 - \bar{\varsigma}_\nu \varsigma} \frac{|\varsigma_\nu|}{\varsigma_\nu} \right) p_\nu \right]$$

be a given product of finitely many binomial J-contractive factors of full rank, and let

$$W_\nu = \mathfrak{A}_\nu^{-1}(0) J \mathfrak{A}_\nu^{*-1}(0) = \begin{bmatrix} -R_\nu & S_\nu^* \\ S_\nu & -T_\nu \end{bmatrix}.$$

Define successively the matrices $\mathfrak{Y}_1$, $\hat{r}_2$, u_2, v_2, $\mathfrak{Y}_2$, $\hat{r}_3$, u_3, v_3, $\mathfrak{Y}_3,\ldots,\mathfrak{Y}_{k-1}$, $\hat{r}_k$, u_k, v_k *with the help of the following recursion process:*

$$\mathfrak{Y}_1 = \begin{bmatrix} \frac{1}{\sqrt{2}}I & \frac{1}{\sqrt{2}}I \\ -\frac{1}{\sqrt{2}}I & \frac{1}{\sqrt{2}}I \end{bmatrix} \begin{bmatrix} R_1^{-1/2} & 0 \\ 0 & |\varsigma_1|^{-1}R_1^{1/2} \end{bmatrix} \begin{bmatrix} I & 0 \\ S_1 R_1^{-1} & I \end{bmatrix} \mathfrak{A}_1^{-1}(0),$$

$$\begin{bmatrix} \frac{1}{\sqrt{2}}I & -\frac{1}{\sqrt{2}}I \\ \frac{1}{\sqrt{2}}I & \frac{1}{\sqrt{2}}I \end{bmatrix} \mathfrak{Y}_1 p_2 \mathfrak{Y}_1^{-1} \begin{bmatrix} \frac{1}{\sqrt{2}}I & \frac{1}{\sqrt{2}}I \\ -\frac{1}{\sqrt{2}}I & \frac{1}{\sqrt{2}}I \end{bmatrix} = \begin{bmatrix} \xi_2^* \\ \eta_2^* \end{bmatrix} [\xi_2, \eta_2] j,$$

$$\hat{r}_2 = I + (1 - |\varsigma_2|^2)|\xi_2|^2,$$

$$\xi_2 = |\xi_2| u_2, \qquad \eta_2 = |\eta_2| v_2;$$

$$\mathfrak{Y}_2 = \begin{bmatrix} \frac{1}{\sqrt{2}}I & \frac{1}{\sqrt{2}}I \\ -\frac{1}{\sqrt{2}}I & \frac{1}{\sqrt{2}}I \end{bmatrix} \begin{bmatrix} \hat{r}_2^{-1/2} u_2 R_1^{-1/2} & 0 \\ 0 & |\varsigma_1|^{-1}|\varsigma_2|^{-1}\hat{r}_2^{1/2} v_2 R_1^{1/2} \end{bmatrix} \begin{bmatrix} I & 0 \\ S_2 R_2^{-1} & I \end{bmatrix} \mathfrak{A}_2^{-1}(0),$$

$$\begin{bmatrix} \frac{1}{\sqrt{2}}I & -\frac{1}{\sqrt{2}}I \\ \frac{1}{\sqrt{2}}I & \frac{1}{\sqrt{2}}I \end{bmatrix} \mathfrak{Y}_2 p_3 \mathfrak{Y}_2^{-1} \begin{bmatrix} \frac{1}{\sqrt{2}}I & \frac{1}{\sqrt{2}}I \\ -\frac{1}{\sqrt{2}}I & \frac{1}{\sqrt{2}}I \end{bmatrix} = \begin{bmatrix} \xi_3^* \\ \eta_3^* \end{bmatrix} [\xi_3, \eta_3] j,$$

$$\hat{r}_3 = I + (1 - |\varsigma_3|^2)|\xi_3|^2,$$

$$\xi_3 = |\xi_3| u_3, \qquad \eta_3 = |\eta_3| v_3;$$

$$\cdots\cdots\cdots\cdots\cdots\cdots\cdots\cdots\cdots\cdots\cdots\cdots\cdots\cdots\cdots$$

$$\mathfrak{Y}_\nu = \begin{bmatrix} \frac{1}{\sqrt{2}}I & \frac{1}{\sqrt{2}}I \\ -\frac{1}{\sqrt{2}}I & \frac{1}{\sqrt{2}}I \end{bmatrix}$$

$$\times \begin{bmatrix} \hat{r}_\nu^{-1/2} u_\nu \hat{r}_{\nu-1}^{-1/2} u_{\nu-1} \cdots \hat{r}_2^{-1/2} u_2 R_1^{-1/2} & \\ 0 & \frac{1}{|\varsigma_1|\cdots|\varsigma_\nu|}\hat{r}_\nu^{1/2} v_\nu \hat{r}_{\nu-1}^{1/2} \cdots \hat{r}_2^{1/2} v_2 R_1^{1/2} \end{bmatrix}$$

$$\times \begin{bmatrix} I & 0 \\ S_\nu R_\nu^{-1} & I \end{bmatrix} \mathfrak{A}_\nu^{-1}(0),$$

$$\begin{bmatrix} \frac{1}{\sqrt{2}}I & -\frac{1}{\sqrt{2}}I \\ \frac{1}{\sqrt{2}}I & \frac{1}{\sqrt{2}}I \end{bmatrix} \mathfrak{Y}_\nu p_{\nu+1} \mathfrak{Y}_\nu^{-1} \begin{bmatrix} \frac{1}{\sqrt{2}}I & \frac{1}{\sqrt{2}}I \\ -\frac{1}{\sqrt{2}}I & \frac{1}{\sqrt{2}}I \end{bmatrix} = \begin{bmatrix} \xi_{\nu+1}^* \\ \eta_{\nu+1}^* \end{bmatrix} [\xi_{\nu+1}, \eta_{\nu+1}] j,$$

$$\hat{r}_{\nu+1} = I + (1 - |\varsigma_{\nu+1}|^2)|\xi_{\nu+1}|^2,$$

$$\xi_{\nu+1} = |\xi_{\nu+1}| u_{\nu+1}, \qquad \eta_{\nu+1} = |\eta_{\nu+1}| v_{\nu+1}$$

$$\cdots\cdots\cdots\cdots\cdots\cdots\cdots\cdots\cdots\cdots\cdots\cdots\cdots\cdots\cdots$$

Then the radii of the Weyl disk are computed by the formulas

$$\rho_{dk} = R_1^{-1/2} u_2^* \hat{r}_2^{-1/2} u_3^* \hat{r}_3^{-1/2} \cdots u_k^* \hat{r}_k^{-1} u_k \cdots \hat{r}_3^{-1/2} u_3 \hat{r}_2^{-1/2} u_2 R_1^{-1/2} \tag{27}$$

$$\rho_{gk} = |\varsigma_1|^2 |\varsigma_2|^2 \cdots |\varsigma_k|^2 R_1^{-1/2} v_2^* \hat{r}_2^{-1/2} v_3^* \hat{r}_3^{-1/2}$$
$$\cdots v_k^* \hat{r}_k^{-1} v_k \cdots \hat{r}_3^{-1/2} v_3 \hat{r}_2^{-1/2} v_2 R_1^{-1/2}. \tag{28}$$

The converse theorem plays an essential role in the construction of an example which is important for us.

THEOREM 5. *Let $R_1 > 0$ and $\hat{r}_2,\ldots,\hat{r}_k \geq I$ be Hermitian matrices, let u_2, $v_2,\ldots,u_k$, v_k be unitary matrices, and let $\varsigma_1,\ldots,\varsigma_k$ be arbitrary distinct points lying inside the unit disk.*

Then there exists a Nevanlinna-Pick problem on the nodes of interpolation at the points $\varsigma_1,\ldots,\varsigma_k$ for which the right and left radii of the Weyl disk at the point $\varsigma = 0$ are determined by (27) and (28).

PROOF. As follows from the remark after Theorem 1, from the given matrix R_1^{-1} it is possible to construct a family of binomial elementary factors such that

R_1^{-1} and $|\varsigma_1|^2 R_1^{-1}$ are the right and left radii of the Weyl disk, respectively. Choose one of them and denote it by $f^{-1}(\varsigma)$. The matrix $\mathfrak{Y}_1$ is defined by (22).

In view of Theorem 3 it is then possible to construct a family of elementary j-contractive factors such that the radii of the Weyl disk are the matrices $u_2^* \hat{r}_2^{-1} u_2$ and $|\varsigma_2|^2 v_2^* \hat{r}_2^{-1} v_2$.

We choose a factor $\beta_2^{-1}(\varsigma)$ in this family and denote by $f_2^{-1}(\varsigma)$ the matrix-valued function

$$f_2^{-1}(\varsigma) = \mathfrak{Y}_1^{-1} \begin{bmatrix} \frac{1}{\sqrt{2}}I & \frac{1}{\sqrt{2}}I \\ -\frac{1}{\sqrt{2}}I & \frac{1}{\sqrt{2}}I \end{bmatrix} \beta_2^{-1}(\varsigma) \begin{bmatrix} \frac{1}{\sqrt{2}}I & -\frac{1}{\sqrt{2}}I \\ \frac{1}{\sqrt{2}}I & \frac{1}{\sqrt{2}}I \end{bmatrix} \mathfrak{Y}_1.$$

Next, let $\mathfrak{A}_2^{-1}(\varsigma) = f_1^{-1}(\varsigma) f_2^{-1}(\varsigma)$, and define the matrix $\mathfrak{Y}_2$ by

$$\mathfrak{Y}_2 = \begin{bmatrix} \frac{1}{\sqrt{2}}I & \frac{1}{\sqrt{2}}I \\ -\frac{1}{\sqrt{2}}I & \frac{1}{\sqrt{2}}I \end{bmatrix} \begin{bmatrix} \hat{r}_2^{-1/2} u_2 R_1^{-1/2} & 0 \\ 0 & \frac{1}{|\varsigma_1||\varsigma_2|} \hat{r}_2^{1/2} v_2 R_1^{1/2} \end{bmatrix}$$
$$\times \begin{bmatrix} I & 0 \\ S_2 R_2^{-1} & I \end{bmatrix} \mathfrak{A}_2^{-1}(0).$$

Again by Theorem 3, we construct a j-contractive factor $\beta_3^{-1}(\varsigma)$ for which the radii of the Weyl disk are the matrices $u_3^* \hat{r}_3^{-1} u_3$ and $|\varsigma_3|^2 v_3^* \hat{r}_3^{-1} v_3$, and we define

$$f_3^{-1}(\varsigma) = \mathfrak{Y}_2^{-1} \begin{bmatrix} \frac{1}{\sqrt{2}}I & \frac{1}{\sqrt{2}}I \\ -\frac{1}{\sqrt{2}}I & \frac{1}{\sqrt{2}}I \end{bmatrix} \beta_3^{-1}(\varsigma) \begin{bmatrix} \frac{1}{\sqrt{2}}I & -\frac{1}{\sqrt{2}}I \\ \frac{1}{\sqrt{2}}I & \frac{1}{\sqrt{2}}I \end{bmatrix} \mathfrak{Y}_2.$$

Continuing this process, we arrive at the matrix-valued function

$$\mathfrak{A}_k^{-1}(\varsigma) = f_1^{-1}(\varsigma) f_2^{-1}(\varsigma) \cdots f_k^{-1}(\varsigma).$$

It follows from Theorem 4 that the radii of the Weyl disk for $\mathfrak{A}_k^{-1}(\varsigma)$ are determined by (27) and (28).

To any product $\mathfrak{A}_k(\varsigma) = f_k(\varsigma) f_{k-1}(\varsigma) \cdots f_1(\varsigma)$ of elementary J-expansive factors of full rank there corresponds [1] a Nevanlinna-Pick problem with interpolation nodes at the poles $\varsigma_1, \ldots, \varsigma_k$.

§3. It will now be assumed that the number of factors in the product

$$\mathfrak{A}_k(\varsigma) = \overset{k}{\underset{\nu=1}{\wedge \!\!\! \prod}} f_\nu(\varsigma)$$

increases without limit. To the sequence $\{\mathfrak{A}_k(\varsigma)\}$ of matrix-valued functions there corresponds a nested sequence of matrix disks $\mathcal{R}_k(\varsigma)$ whose right and left radii ρ_{d_k} and ρ_{g_k} can be computed by (27) and (28).

The limit of the centers of the disks $\mathcal{R}_k$ exists, because a computation of the J-form of a group parametrized factor of second order already shows that the condition $L > 0$ holds.

Consequently, we can now speak of the limit Weyl disk at each point ς_0 of the unit disk. By a theorem of Orlov [4], the ranks of the right and left limit radii do not depend on the choice of the point ς_0, and this makes it possible to classify Nevanlinna-Pick problems according to the values of the ranks of the limit radii. The structural formulas (27) and (28) play an essential role in these questions. In particular, they have the following corollary.

COROLLARY. *For each Nevanlinna-Pick problem with right limit radius of full rank and such that the scalar Blaschke product $\prod_1^\infty |\varsigma_\nu|$ converges, the left limit radius $\rho_{g\infty}$ also has full rank; but if $\prod_1^\infty |\varsigma_\nu|$ diverges, then the left limit radius $\rho_{g\infty}$ is equal to zero.*

Obviously, if the left limit radius has full rank, then $\prod_1^\infty |\varsigma_\nu|$ converges, and the right limit radius has full rank.

In the case when the limit radii are singular matrices, there is no connection between their ranks. Namely, for any pair of nonnegative integers ν_d and ν_g, each less than m, it is possible to construct a Nevanlinna-Pick problem such that the ranks of the radii of the limit Weyl disk are equal to ν_d and ν_g, respectively. The construction is based on Theorem 5.

Let $R_1 = I$, and let $\hat{r}_2, \ldots, \hat{r}_k, \ldots$ be diagonal matrices with elements ≥ 1 and arranged in decreasing order. For definiteness assume that $\nu_d \geq \nu_g$. Let $u_2 = u_3 = \cdots = u_k = \cdots = I$ and

$$
\hat{r}_k^{-1} = \begin{bmatrix} \gamma_1^{(k)} & & & & & & & \\ & \gamma_2^{(k)} & & & & & & \\ & & \ddots & & & & & \\ & & & \gamma_{\mu_d}^{(k)} & & & & \\ & & & & \delta_1^{(k)} & & & \\ & & & & & \ddots & & \\ & & & & & & \delta_{\nu_d}^{(k)} \end{bmatrix}, \qquad \mu_d + \nu_d = m
$$

and choose sequences $\{\gamma_i^{(k)}\}$ $(i = 1, 2, \ldots, \mu_d)$ such that the products $\prod_{k=2}^\infty \gamma_i^{(k)}$ diverge (i.e., tend to zero), and sequences $\{\delta_j^{(k)}\}$ $(j = 1, 2, \ldots, \nu_d)$ such that the products $\prod_{k=2}^\infty \delta_j^{(k)}$ converge (i.e., tend to a nonzero limit).

For this choice of the matrices R_1, $\hat{r}_k$, and u_k the right radius ρ_{dn} has the form

$$
\rho_{dn} = \begin{bmatrix} \prod_{k=2}^n \gamma_1^{(k)} & & & & & & \\ & \prod_{k=2}^n \gamma_2^{(k)} & & & & & \\ & & \ddots & & & & \\ & & & \prod_{k=2}^n \gamma_{\mu_d}^{(k)} & & & \\ & & & & \prod_{k=2}^n \delta_1^{(k)} & & \\ & & & & & \ddots & \\ & & & & & & \prod_{k=2}^n \delta_{\nu_d}^{(k)} \end{bmatrix},
$$

and the right radius

$$
\rho_{d\infty} =
\begin{bmatrix}
0 & & & & & & & \\
& 0 & & & & & & \\
& & \ddots & & & & & \\
& & & 0 & & & & \\
& & & & \Delta_1 & & & \\
& & & & & \ddots & & \\
& & & & & & \Delta_{\nu_d}
\end{bmatrix}
\qquad (\Delta_j > 0, \ j = 1, 2, \ldots, \nu_d)
$$

of the limit Weyl disk has rank equal to ν_d.

To construct the left radius we should now define the unitary matrices $v_2, v_3, \ldots$. We specify them in such a way that ρ_g has diagonal form, with the first $\mu_g = m - \nu_g$ diagonal elements of it tending to zero. This is accomplished by transposing the first diagonal element in some of the $\hat{r}_k^{-1}$ with the diagonal elements in the $\mu_d + 1$, $\mu_d + 2, \ldots, \mu_g$ positions.

For example, transposition of the first diagonal element with the one at the position $\mu_d + 1$ is implemented by means of the transformation $v^* \hat{r}^{-1} v$, where the matrix v^* has the form

$$
v^* =
\left[
\begin{array}{ccccc|ccccc}
0 & 0 & \cdots & 0 & 1 & 0 & \cdots & 0 \\
0 & 1 & \cdots & 0 & 0 & 0 & \cdots & 0 \\
\multicolumn{9}{c}{\dotfill} \\
0 & 0 & \cdots & 1 & 0 & 0 & \cdots & 0 \\
\hline
1 & 0 & \cdots & 0 & 0 & 0 & \cdots & 0 \\
0 & 0 & \cdots & 0 & 0 & 1 & \cdots & 0 \\
\multicolumn{9}{c}{\dotfill} \\
0 & 0 & \cdots & 0 & 0 & 0 & \cdots & 1
\end{array}
\right], \qquad v^* v = I.
$$

The transposition of the diagonal elements that ensures zero divergence is carried out as follows. Let $\nu_d - \nu_g = \nu - 1$. The series

$$
\sum_{k=2}^{\infty} (1 - \gamma_1^{(k)})
$$

diverges. We break it up into segments each of which is greater than 1. The elements of the matrices $\hat{r}_k^{-1}$ corresponding to the segments with numbers 1, $\nu + 1$, $2\nu + 1$, $3\nu + 1, \ldots$ are left in place. The elements corresponding to the segments with numbers 2, $\nu + 2$, $2\nu + 2$, $3\nu + 2, \ldots$ are transposed with the elements in the position $\mu_d + 1$; the elements of the matrices $\hat{r}_k^{-1}$ corresponding to the segments with numbers 3, $\nu + 3$, $2\nu + 3$, $3\nu + 3, \ldots$ are transposed with the elements in the position $\mu_d + 2$, and so on.

It is clear that after such a rearrangement the μ_g first series $\sum (1 - h^{(k)})$ now diverge; hence so do the corresponding products; the left radius of the limit Weyl disk has the form

$$
\rho_{g\infty} =
\begin{bmatrix}
0 & & & & & & \\
& 0 & & & & & \\
& & \ddots & & & & \\
& & & 0 & & & \\
& & & & \Delta_\nu & & \\
& & & & & \ddots & \\
& & & & & & \Delta_{\nu_d}
\end{bmatrix}.
$$

By Theorem 5, to such a sequence of matrices $R_1 = I$, $\hat{r}_k$, $u_k = I$, and v_k there corresponds an infinite product

$$
\cdots f_k(\varsigma) f_{k-1}(\varsigma) \cdots f_2(\varsigma) f_1(\varsigma)
$$

of binomial J-expansive factors of full rank, and hence also a Nevanlinna-Pick problem connected with it.

BIBLIOGRAPHY

1. Rolf Nevanlinna, *Über beschränkte analytische Funktionen*, Ann. Acad. Sci. Fenn. Ser. A **32** (1929), no. 7.

2. A. V. Efimov and V. P. Potapov, *J-expansive matrix-valued functions and their role in the analytic theory of electrical circuits*, Uspekhi Mat. Nauk **28** (1973), no. 1(169), 65–130; English transl. in Russian Math. Surveys **28** (1973).

3. I. V. Kovalishina and V. P. Potapov, *An indefinite metric in the Nevanlinna-Pick problem*, Akad. Nauk Armyan. SSR Dokl. **59** (1974), 17–22; English transl. in this volume.

4. S. A. Orlov, *Nested matrix disks depending analytically on a parameter, and theorems on the invariance of ranks of radii of limiting disks*, Izv. Akad. Nauk SSSR Ser. Mat. **40** (1976), 593–644; English transl. in Math. USSR Izv. **10** (1976).

Translated by H. H. MCFADEN

Amer. Math. Soc. Transl.
(2) Vol. **138**, 1988

A Theorem on the Modulus. I
Main Concepts. The Modulus

UDC 519.210

V. P. POTAPOV

The matrices considered here are initially interpreted as operators acting in an m-dimensional vector space with an indefinite metric. The concept of the modulus of a J-nonexpansive matrix is introduced. A basic theorem on a product of moduli is proved.

§1. Preliminary facts

The main objects to be studied here are $m \times m$ matrices $A = (a_{kj})$ with complex elements. Along with them we consider $k \times m$ matrices and, in particular, matrices $f = [\xi_1, \ldots, \xi_m]$, also called vectors. The matrix $A^* = \overline{A}'$, where the prime denotes passage to the transposed matrix and the bar denotes passage to the complex conjugate elements, is called the *adjoint* of a matrix A.

The following properties of the adjoint operation are well known: 1. $(A^*)^* = A$; 2. $(A + B)^* = A^* + B^*$; 3. $(\lambda A)^* = \overline{\lambda} A^*$ (λ a scalar); 4. $(AB)^* = B^* A^*$.

Applied to a vector f, the adjoint operation leads to a column matrix f^* of dimension $m \times 1$. In this connection the matrix product $fg^* = \sum_1^m \xi_k \overline{\eta}_k = \langle f, g \rangle$ is a number called the *inner product* of f and g, while $f^* g = (\xi_k \overline{\eta}_k)$ is an $m \times m$ matrix of rank 1.

A matrix H is said to be *Hermitian* if $H^* = H$. A Hermitian matrix H is said to be *positive* if the *Hermitian form* $fHf^* = \sum_{k,j=1}^m \xi_k h_{kj} \overline{\xi}_j$ is positive for all vectors $f \neq 0$. If $fHf^* \geq 0$, then H is said to be *nonnegative*. In general, H is said to be *definite* if fHf^* does not change sign for all possible vectors f, and *indefinite* otherwise.

1980 *Mathematics Subject Classification* (1985 *Revision*). Primary 15A57, 47B50, 47H09; Secondary 15A21.

Translation of Teor. Funktsiĭ, Funktsional. Anal. i Prilozhen. Vyp. 38 (1982), 91–101; MR **84d**:15026.

The inequality $A \geq B$ means that $A - B \geq 0$, i.e., $A - B$ is a Hermitian nonnegative matrix.

We shall very often use the fact that if $A \geq B$, then $TAT^* \geq TBT^*$ for any $l \times m$ matrix T (l any positive integer). The converse is true for every nonsingular square matrix T.

A matrix U is said to be *unitary* if $UU^* = I$, and a matrix X is said to be *nonexpansive* (a contraction) if $XX^* \leq I$.

Let g be an arbitrary vector and let $gX = f$. Then

$$\langle f, f \rangle = \langle gX, gX \rangle = gXX^*g^* \leq gIg^* = gg^* = \langle g, g \rangle,$$

and hence a contraction X carries each vector g into a vector $f = gX$ which is not longer. Similarly, a unitary matrix U carries a vector g into a vector $f = Ug$ of the same length. The fact that the adjoint U^* is unitary when U is can be seen directly from the definition. The analogous assertion for contractions X can be proved most simply by using the concept of the operator norm of a matrix (operator) A:

$$\|A\| = \sup_f (\|fA\|/\|f\|),$$

where the number $\|f\| = \langle f, f \rangle^{1/2} = (\sum |\xi_j|^2)^{1/2}$ is taken as the norm of a vector. It is easy to prove that

$$\|A\| = \sup_{f,g} \frac{|\langle fA, g \rangle|}{\|f\| \cdot \|g\|}.$$

And since $|\langle fA, g \rangle| = |\langle gA^*, f \rangle|$, it follows that $\|A^*\| = \|A\|$. The condition that X is a contraction, $I - XX^* \geq 0$, is equivalent to the inequality $\|X\| < 1$. But then also $\|X^*\| < 1$, and hence X^* is a contraction: $I - X^*X \geq 0$.

The concept of the modulus of a matrix A is introduced by analogy with the modulus of a complex number a. The number $r = (a\bar{a})^{1/2}$ is defined to be the modulus of the number a, and, since $(r^{-1}a)\overline{(r^{-1}a)} = 1$, it follows that $r^{-1}a = e^{i\varphi}$, which yields the polar representation $a = re^{i\varphi}$.

Suppose now that A is an arbitrary square matrix. The product $AA^* = H$ is a Hermitian-nonnegative matrix. By a unitary transformation U we reduce it to a diagonal form D: $H = UDU^*$. Since $D \geq 0$ here, the square root $D^{1/2}$ exists. Setting $R = UD^{1/2}U^*$, we get that $R^2 = AA^*$. Regarding A as invertible for simplicity, we conclude from the equality $R^{-1}(AA^*)R^{-1} = I$ that the matrix $R^{-1}A = w$ is unitary. The equality $A = RW$ is an analogue of the polar representation of a complex number.

In concluding this section we give a brief list of facts we need about reduction of a square matrix to Jordan normal form and the concept of a function of a matrix.

In our notation the expression $gA = \lambda g$, $g \neq 0$, means that g is an eigenvector of A corresponding to the eigenvalue λ.

Each square matrix A can be reduced to Jordan normal form C by a similarity transformation $A = TCT^{-1}$. The matrix C is a block-diagonal matrix. Its diagonal blocks, called Jordan cells, have the form

$$
C(\lambda_0) = \begin{pmatrix}
\lambda_0 & 1 & 0 & \cdots & 0 & 0 \\
0 & \lambda_0 & 1 & \cdots & 0 & 0 \\
\cdots & \cdots & \cdots & \cdots & \cdots & \cdots \\
0 & 0 & 0 & \cdots & \lambda_0 & 1 \\
0 & 0 & 0 & \cdots & 0 & \lambda_0
\end{pmatrix};
$$

here λ_0 is an eigenvalue, and the order of the cell is equal to m_0. Several Jordan cells can corrrespond to the same eigenvalue λ_0, with the order of the cells not necessarily the same. Associated with each Jordan cell is the binomial $(\lambda-\lambda_0)^{m_0}$, called an elementary divisor of A. The collection of all elementary divisors is an invariant of a similarity transformation, and it determines the structure of the Jordan matrix C to within a permutation of the cells.

All the elementary divisors of a Hermitian matrix H are simple, i.e., have the form $\lambda - \lambda_0$. In other words, the Jordan form of a Hermitian matrix H is a diagonal matrix.

There are several ways to define a function of a matrix. They all lead to the same final result. We dwell on three of them.

The first, and perhaps the most general, was presented by F. Riesz. Let $\sigma(A)$ denote the spectrum of A: the collection of its eigenvalues. If $f(z)$ is a holomorphic function on a domain G containing $\sigma(A)$, and Γ is a contour in G about $\sigma(A)$, then, in correspondence to the Cauchy formula

$$
f(z) = \frac{1}{2\pi i} \oint_\Gamma \frac{f(\varsigma)}{\varsigma - z} \, d\varsigma,
$$

we set

$$
f(A) = \frac{1}{2\pi i} \oint_\Gamma (\varsigma I - A)^{-1} f(\varsigma) \, d\varsigma.
$$

A merit of the Riesz definition is that it also encompasses the case of operators acting in an infinite-dimensional space, and a deficiency is that the construction has a certain diffusiveness.

The second way can be used when $f(z)$ is holomorphic in a disk $|z| < R$ containing $\sigma(A)$. Starting from the power series expansion $f(z) = a_0 + a_1 z + a_2 z^2 + \ldots$ of $f(z)$, we let $f(A) = a_0 + a_1 A + a_2 A^2 + \cdots$. This is trivially a good way when $f(z)$ is an entire function.

Finally, the third way is to reduce A to Jordan normal form $A = TCT^{-1}$ and replace each cell $C(\lambda_0)$ by the cell

$$
\begin{pmatrix}
f(\lambda_0) & \dfrac{f'(\lambda_0)}{1!} & \cdots & \dfrac{f^{m_0-1}(\lambda_0)}{(m_0 - 1)!} \\
0 & f(\lambda_0) & \cdots & \dfrac{f^{m_0-2}(\lambda_0)}{(m_0 - 2)!} \\
\cdots & \cdots & \cdots & \cdots \\
0 & 0 & \cdots & f(\lambda_0)
\end{pmatrix} = f[C(\lambda_0)].
$$

Then $f(A)$ is defined by the equality $f(A) = Tf(C)T^{-1}$. It can be proved that the result does not depend on the choice of the matrix T reducing A to Jordan normal form.

§2. Main concepts of a J-algebra

Let $J = (P_{kj})$ be a Hermitian indefinite (in general) $m \times m$ matrix whose square is the identity: $J^* = J$ and $J^2 = I$. We introduce an *indefinite metric* in the m-dimensional complex linear space L of row vectors $f = [\xi_1, \ldots, \xi_m]$ by defining the inner product

$$[f, g] = fJg^* = \sum_{k,j=1}^{m} \xi_k P_{kj} \overline{\eta}_j.$$

As in the usual case of the metric $\langle f, g \rangle = fg^* = \sum_i^m \xi_k \overline{\eta}_k$, the bilinear functional $[f, g]$ has the following properties:

1. $[\lambda_1 f_1 + \lambda_2 f_2, g] = \lambda_1 [f_1, g] + \lambda_2 [f_2, g]$;
2. $[g, f] = \overline{[f, g]}$.

However, it cannot now be asserted that the scalar square $[f, f]$ of a vector $f \neq 0$ is a positive number. In this connection the vectors $g \in L$ are divided into "plus-vectors", "minus-vectors", and "neutral vectors", according to whether $[g, g]$ is positive, negative, or zero. In the general theory the form of the metrizing matrix J is inessential. However, in concrete situations the following forms of J are most often encountered (the Pauli matrices):

$$J = \begin{pmatrix} -I_q & 0 \\ 0 & I_p \end{pmatrix} \quad (p + q = m); \qquad J_r = \begin{pmatrix} 0 & I \\ I & 0 \end{pmatrix}; \qquad J_s = \begin{pmatrix} 0 & -iI \\ iI & 0 \end{pmatrix}.$$

The matrix J_r is closely connected with the theory of electric fields, and J_s is closely connected with classical problems; in these two cases m must be an even integer.

The presence of an inner product enables us to assign to each matrix the *J-adjoint matrix A^+* by requiring that $[f, gA^+] = [fA, g]$ for all f and g in L. This relation means that $fJ(A^+)^*g^* = fAJg^*$, which, since f and g are arbitrary vectors and $J^2 = I$, implies that $A^+ = JA^*J$.

A matrix H is said to be *J-Hermitian (J-selfadjoint)* if $H = H^+$, or, equivalently, $HJ = JH^*$.([1]) Obviously, for a J-Hermitian matrix the inner product $[fH, f]$ is a real number for all $f \in L$. But if in addition this inner product is nonnegative for all $f \in L$, then H is called a *J-Hermitian-nonnegative* matrix, and clearly is characterized by the inequality $HJ \geq 0$.

A matrix U is said to be *J-unitary* if it preserves the scalar square: $[fU, fU] = [f, f]$ $(f \in L)$. Since equality of quadratic forms on a complex vector space implies equality of the corresponding bilinear forms, it follows that $[fU, gU] = [f, g]$, and hence $UJU^* = J$.

([1])The asterisk is dropped when H^* is commuted with J.

Finally, a matrix W is said to be *J-nonexpansive* if it does not increase the scalar square: $[fW, fW] \leq [f, f]$, i.e., if $WJW^* \leq J$. The expression $J - WJW^* \geq 0$ (often encountered in what follows) is called a *J-form*.

The theory presented here studies the class $\mathbf{W}$ of J-nonexpansive matrices. A decisive circumstance is the multiplicativity of this class: if $W_1, W_2 \in \mathbf{W}$, then $W_1 W_2 \in \mathbf{W}$. Indeed,

$$J - (W_1 W_2) J (W_1 W_2)^* = J - W_1 J W_1^* + W_1 (J - W_2 J W_2^*) W_1^* \geq 0.$$

The following result is fundamental.

THEOREM 2.1. *If W is J-nonexpansive (J-unitary), then so is W^*.*

For a proof we consider a fractional linear transformation of the class $\mathbf{W}(W)$ into the class $\mathbf{X}(X)$ of contractions (relations of this kind are often encountered in other questions in J-theory). To realize this connection we introduce the matrices $P = \frac{1}{2}(I + J)$, $Q = \frac{1}{2}(I - J)$. It is easy to see that P and Q are orthogonal projections, $P^2 = P = P^* \geq 0$, $Q^2 = Q = Q^* \geq 0$, $PQ = QP = 0$, $P + Q = I$, and $P - Q = J$.

Three assertions hold for the fractional linear transformation

$$X = (WQ - P)^{-1}(-WP + Q). \tag{2.1}$$

First, this transformation makes sense for any J-nonexpansive matrix W; second, it can be written in the form

$$X = (PW + Q)(QW + P)^{-1}; \tag{2.2}$$

third, it carries J-nonexpansive matrices into contractions X, and J-unitary matrices into unitary matrices.

1. The condition that W is J-nonexpansive is clearly equivalent to the inequality $P + WQW^* \geq Q + WPW^*$. Multiplying the denominator in (2.1) by the adjoint expression from the right, we get that

$$(WQ - P)(QW^* - P) = WQW^* + P.$$

The last two relations give us that if $g(WQ - P) = 0$, then $gP = 0$ and $gQ = 0$ simultaneously, i.e., $g = 0$.

2. It is easy to verify the identity

$$(WQ - P)(PW + Q) = (-WP + Q)(QW + P), \tag{2.3}$$

which implies that $QW + P$ is a nonsingular matrix. Indeeed, if $(QW + P)h^* = 0$, then it follows from (2.3) and 1 that $(PW + Q)h^* = 0$. Multiplying the first by P from the left and the second by Q from the left and adding, we get that $h^* = 0$. In view of this identity, (2.3) can be rewritten in the form

$$(WQ - P)^{-1}(-WP + Q) = (PW + Q)(QW + P)^{-1}.$$

3. Using the first and second transformation formulas, we compute

$$I - XX^* = (WQ - P)^{-1}\{J - WJW^*\}(QW^* - P)^{-1},$$
$$I - X^*X = (W^*Q + P)^{-1}\{J - W^*JW\}(QW + P)^{-1}.$$

After this, the assertions of the theorem become consequences of known properties of contractions and unitary matrices.

§3. Definition of the modulus of a J-nonexpansive matrix

The concept in §1 of the modulus of an arbitrary matrix A was based on the construction of a matrix R with nonnegative eigenvalues and with square equal to AA^*.

The following hold for each matrix S with positive eigenvalues:

1) There exists a unique matrix R (called the *square root of S* and denoted by $R = S^{1/2}$) with positive eigenvalues such that $R^2 = S$.

2) There exists a unique matrix H (called the *principal value of the logarithm of S* and denoted by $H = \ln S$) with real eigenvalues such that $e^H = S$.

Note that for Hermitian matrices $S \geq 0$ the square root $S^{1/2}$ makes sense also when there are eigenvalues equal to zero. In the general case this is not so because of the possible presence in S of Jordan cells of the form $s = \begin{pmatrix} 0 & 0 \\ 1 & 0 \end{pmatrix}$, for which $s^{1/2}$ does not exist.

In generalizing the concept of the modulus to the case of an indefinite metric it is natural to consider the matrix $S = AA^+ = AJA^*J$ instead of AA^*.

However, we take into account that for an arbitrary matrix A the usual procedure for constructing the modulus cannot always be realized, as is clear from the example

$$A = \begin{pmatrix} 0 & 1 \\ 1 & 0 \end{pmatrix}, \qquad J = \begin{pmatrix} -1 & 0 \\ 0 & 1 \end{pmatrix}.$$

Here $S = AA^+ = \begin{pmatrix} -1 & 0 \\ 0 & -1 \end{pmatrix}$ and the square root $S^{1/2}$ does not make sense.

In fact, the problem of constructing a modulus should be posed not for arbitrary matrices A but for matrices W of the class **W**, and also for invertible matrices. For these matrices we have

THEOREM 3.1. *If W is an invertible J-nonexpansive matrix, then all the eigenvalues of the matrix $S = WJW^*J$ are positive.*

PROOF. Let $H = I - S$. Since $HJ = J - WJW^*$, it follows that H is a J-Hermitian nonnegative matrix.

1) First we show that the spectrum $\sigma(H)$ is real. Indeed, let $g \neq 0$ be an eigenvector of H corresponding to the eigenvalue λ: $\lambda g = gH$. Then $\lambda g J g^* = gHJg^*$. If $gHJg^* > 0$, then λ is real, but if $gHJg^* = 0$, then we get successively that $g(HJ)^{1/2} \cdot (HJ)^{1/2}g^* = 0$, $g(HJ)^{1/2} = 0$, $g(HJ)^{1/2}(HJ)^{1/2} = 0$, $g(HJ) = 0$, $gH = 0$, $\lambda g = 0$, and, consequently, $\lambda = 0$.

2) We now show that $\sigma(S)$ is positive. Indeed, let $e \neq 0$ be an eigenvector of S corresponding to the eigenvalue μ: $\mu e = eS = e(I - H)$. Then $(1 - \mu)e = eH$, which implies that

$$(1 - \mu)eJe^* = eHJe^*. \tag{3.1}$$

Let us now use the fact that W^* is J-nonexpansive simultaneously with W: $J - W^*JW \geq 0$. Hence

$$WJ \cdot J \cdot JW^* \geq WJW^*J \cdot J \cdot JWJW^*$$

or $SJ \geq SJS^*$. But then $eSJe^* \geq eSJS^*e^*$.

Since μ is real by what was proved in 1), it follows that $\mu eJe^* \geq \mu^2 eJe^*$ or $\mu(1 - \mu)eJe^* \geq 0$. Using (3.1), we arrive at the relation $\mu eHJe^* \geq 0$. If $eHJe^* > 0$, then $\mu \geq 0$, but if $eHJe^* = 0$, then, as in 1), $eH = 0$, $1 - \mu = 0$, and hence $\mu = 1$. Thus the eigenvalues of S are nonnegative, and, since S is invertible, they are positive. This proves the theorem.

REMARK. The spectral properties of the J-Hermitian-nonnegative matrix H (and hence also of S) can be made more precise without difficulty. The elementary divisors of H corresponding to eigenvalues $\lambda \neq 0$ have multiplicity 1, while those corresponding to the eigenvalue $\lambda = 0$ have only multiplicity 2. An exhaustive analysis of the Jordan structure of any (not necessarily nonnegative) J-Hermitian matrix was given by Frobenius; a presentation of his results can be found in [1].

COROLLARY. *For any invertible J-nonexpansive matrix W there exists a matrix R such that*

$$R = S^{1/2} = (WW^+)^{1/2} = (WJW^*J)^{1/2}.$$

*The matrix R is called the **modulus** of W.*

§4. Simplest properties of the modulus

A number of properties of the matrix $S = WJW^*J$ carry over to the modulus R of a $W \in \mathbf{W}$. To derive them it is convenient to use the connection between an arbitrary matrix A with positive spectrum and its square A^2, expressed by the formula

$$A^{-1} = \frac{2}{\pi} \int_0^{+\infty} (x^2 I + A^2)^{-1}\, dx. \tag{4.1}$$

This follows from the corresponding trivial scalar relation

$$\int_0^{+\infty} \frac{dx}{x^2 + a^2} = \frac{\pi}{2} a^{-1} \qquad (a > 0)$$

and is easily verified on the Jordan cells of A.

The meaning of (4.1) is that it reduces the study of the relatively complicated dependence of A on A^2 to the study of a "linear combination" of the functions $(x^2 I + A^2)^{-1}$ with positive coefficients, and this is of a considerably simpler nature.

The next result is obvious.

LEMMA 4.1. *If A is an invertible J-Hermitian matrix, then A^{-1} is also J-Hermitian.*

The relations $AJ = JA^*$ and $JA^{-1*} = A^{-1}J$ are equivalent, since $(A^{-1})^* = (A^*)^{-1}$.

From Lemma 4.1 and (4.1) we get

LEMMA 4.2. *If S is a J-Hermitian matrix with positive eigenvalues, then $S^{1/2}$ is also J-Hermitian.*

We have that

$$S^{-1/2} = \frac{2}{\pi} \int_0^{+\infty} (x^2 I + S)^{-1} \, dx,$$

and the fact that S is J-Hermitian implies the same property successively for the matrices $x^2 I + S$, $(x^2 I + S)^{-1}$, $S^{-1/2}$, and $S^{1/2}$.

LEMMA 4.3. *If A and B are J-Hermitian matrices with positive eigenvalues, then the inequalities $AJ \leq BJ$ and $A^{-1}J \geq B^{-1}J$ are equivalent.*

Indeed, $A^{1/2}$ and $B^{1/2}$ exist, are invertible, and are J-Hermitian by Lemma 4.2. In view of this, the inequality $AJ \leq BJ$ is equivalent to $A^{1/2}J(A^{1/2})^* \leq B^{1/2}J(B^{1/2})^*$, or $B^{-1/2}A^{1/2}J(A^{1/2})^*(B^{-1/2})^* \leq J$, which means that the matrix $W = B^{-1/2}A^{1/2}$ is J-nonexpansive. But then the matrix W^*, $W^*JW \leq J$, is also J-nonexpansive by the theorem in §1, i.e.,

$$(A^{1/2})^*(B^{-1/2})^*JB^{-1/2}A^{1/2} \leq J,$$

which implies that $(B^{-1/2})^*JB^{-1/2} \leq (A^{-1/2})^*JA^{-1/2}$, or $JB^{-1} \leq JA^{-1}$. Multiplying both sides of the inequality from the left and from the right by J, we get that $B^{-1}J \leq A^{-1}J$, which is what we were required to prove.

After this, it is not difficult to prove a theorem establishing the main properties of the modulus R of a J-non-expansive matrix W.

THEOREM 4.1. *The modulus $R = (WJW^*J)^{1/2}$ of an invertible J-non-expansive matrix W is a J-Hermitian J-nonexpansive matrix with positive eigenvalues. There is a "polar representation" of W: $W = RU$, where U is a J-unitary matrix.*

PROOF. The fact that R is J-Hermitian follows on the basis of Lemma 4.2 from the fact that in this case the matrix $S = WJW^*J$ is J-Hermitian. Using this, we get the J-nonexpansiveness of

$$R: J - RJR^* = J - R^2 J = J - SJ = J - WJW^* \geq 0.$$

The positivity of the eigenvalues is contained in the definition $S^{1/2} = R$. Finally, the equality $RJR^* = WJW^*$ implies that $U = R^{-1}W$ is a J-unitary matrix $W = RU$; and this concludes the proof.

The following theorem, which in a certain sense is a converse, gives an intrinsic determination of the modulus.

THEOREM 4.2. *The modulus R of a J-Hermitian J-nonexpansive matrix W with positive eigenvalues coincides with W.*

We have that $R^2 = S = WJW^*J = W \cdot WJ \cdot J = W^2$, and it follows from the uniqueness of the square root of a matrix S with positive eigenvalues that $R = W$.

THEOREM 4.3. *The modulus R satisfies the inequalities*

$$0 \leq J - RJ \leq J - R^2 J, \qquad 0 \leq R^{-1}J - J \leq R^{-2}J - J.$$

PROOF. Starting from (4.1) with A replaced by R, we compute

$$R^{-1}J - J = \left[\frac{2}{\pi}\int_0^{+\infty}(x^2 I + R^2)^{-1}\,dx - \frac{2}{\pi}\int_0^{+\infty}(x^2 I + I)^{-1}\,dx\right]J$$

$$= \frac{2}{\pi}\int_0^{+\infty}(x^2 I + R^2)^{-1}(I - R^2)J\frac{dx}{x^2 + 1}$$

$$= \frac{2}{\pi}\int_0^{+\infty}(x^2 I + R^2)^{-1/2}(I - R^2)(x^2 I + R^2)^{-1/2}J\frac{dx}{x^2 + 1}$$

$$= \frac{2}{\pi}\int_0^{+\infty}(x^2 I + R^2)^{-1/2}(J - RJR^*)[(x^2 I + R^2)^{-1/2}]^*\frac{dx}{x^2 + 1} \geq 0.$$

Here we have used the fact that the matrices R and $(x^2 I + R^2)^{-1/2}$ are J-Hermitian; the latter follows from Lemmas 4.1 and 4.2. Therefore $R^{-1}J \geq I \cdot J$, and so $RJ \leq I \cdot J$ by Lemma 4.3. This proves the inequalities $0 \leq J - RJ$ and $0 \leq R^{-1}J - J$. But

$$J - R^2 J = J - RJ + RJ - R^2 J$$

$$= J - RJ + R^{1/2}J(R^{1/2})^* - RJR^*$$

$$= J - RJ + R^{1/2}(J - RJ)(R^{1/2})^* \geq J - RJ,$$

which concludes the proof.

THEOREM 4.4. *If R_1 and R_2 are moduli, then $R_1^2 J \leq R_2^2 J$ implies $R_1 J \leq R_2 J$.*

PROOF. By (4.1),

$$R_i^{-1}J = \frac{2}{\pi}\int_0^{+\infty}(x^2 I + R_i^2)^{-1}J\,dx \qquad (i = 1, 2).$$

Since $(x^2 I + R_1^2)J \leq (x^2 I + R_2^2)J$ by assumption, it follows from Lemma 4.3 that

$$(x^2 I + R_1^2)^{-1}J \geq (x^2 I + R_2^2)^{-1}J,$$

and hence $R_1^{-1}J \geq R_2^{-1}J$. Using Lemma 4.3 again, we get that $R_1 J \leq R_2 J$, as asserted.

As we know, the converse is not true even in the case of the definite metric $J = I$.

§5. An exponential representation of the modulus

To get a formula for $\ln A$ analogous to the integral representation (4.1) we integrate the equality

$$\frac{1}{a} = \frac{2}{\pi}\int_0^{+\infty}\frac{dx}{x^2 + a^2}$$

with respect to the parameter a from 1 to an $a > 0$:

$$\ln a = \frac{2}{\pi} \int_0^{+\infty} \frac{1}{x} \left[\arctan \frac{a}{x} - \arctan \frac{1}{x} \right] dx.$$

Integrating by parts and considering that the double substitution is equal to zero, we get that

$$\ln a = \frac{2}{\pi} \int_0^{+\infty} \left[\frac{a}{x^2 + a^2} - \frac{1}{x^2 + 1} \right] \ln x \, dx.$$

Breaking up this integral into a sum of integrals over $[0, 1]$ and $[1, +\infty]$, we replace x by $1/x$ in the latter. This gives us finally that

$$\ln a = \frac{2}{\pi} \int_0^1 \left[\frac{a}{x^2 + a^2} - \frac{a}{1 + a^2 x^2} \right] \ln x \, dx.$$

The fact that this equality remains valid also after the number $a > 0$ is replaced by an arbitrary matrix A with positive spectrum can be checked in the same way as formula (4.1) on the Jordan cells of A. The following expression for it is expedient here:

$$\ln A = \frac{2}{\pi} \int_0^1 (x^2 I + A^2)^{-1/2} (I + A^2 x^2)^{-1/2} A^{1/2} \{I - A^2\} A^{1/2}$$

$$\times (I + A^2 x^2)^{-1/2} (x^2 I + A^2)^{-1/2} (1 - x^2) \ln x \, dx.$$

Now let $A = R$ be a modulus, i.e., a J-nonexpansive J-Hermitian matrix with positive spectrum. Let $\ln A = -H$. Considering Lemmas 4.1 and 4.2, we arrive at the following relation:

$$HJ = \frac{2}{\pi} \int_0^1 [(x^2 I + R^2)^{-1/2} (I + R^2 x^2)^{-1/2} R^{1/2} (J - RJR^*)$$

$$\times [(x^2 I + R^2)^{-1/2} (I + R^2 x^2)^{-1/2}] R^{1/2*} (1 - x^2) \ln x \, dx.$$

We thereby get

THEOREM 5.1. *The modulus* $R = (WJW^*J)^{1/2}$ *of an invertible J-non-expansive matrix W can be uniquely represented in the form* $R = e^{-H}$, *where H is a J-Hermitian-nonnegative matrix.*

We also have the converse:

THEOREM 5.2. *If H is an arbitrary J-Hermitian-nonnegative matrix, then the matrix* $R = e^{-H}$ *is a modulus.*

PROOF. Since the eigenvalues of H are real, the spectrum of R is positive. Furthermore, $RJ = J \cdot J^{-1} e^{-H} J = J e^{-JHJ} = J e^{-H^*} = JR^*$, i.e., R is J-Hermitian.

It remains to prove that R is J-nonexpansive. Let g be an arbitrary vector. We consider the scalar function

$$\varphi_g(t) = [g e^{-tH}, g e^{-tH}] = g e^{-tH} J e^{-tH^*} g^*$$

of the real variable t and find its derivative:

$$\varphi_g'(t) = ge^{-tH}(-HJ)e^{-tH^*}g^* + ge^{-tH}J(-H^*)e^{-tH^*}g^*$$
$$= -2ge^{-tH}(HJ)e^{-tH^*}g^* \le 0.$$

Thus, $\varphi_g(t)$ is monotonically decreasing; in particular, $\varphi_g(0) \ge \varphi_g(1)$, i.e., $gJg^* \ge gRJR^*g^*$. Since g was arbitrary, this concludes the proof of the theorem.

Three consequences of these theorems should be mentioned.

1. *If $R = e^{-H}$ is a modulus, then $R^t = e^{-tH}$ is also a modulus for any $t \ge 0$.*

2. Theorem 4.3 follows from the inequality $\varphi_g(-1) \ge \varphi_g(-1/2) \ge \varphi_g(0) \ge \varphi_g(1/2) \ge \varphi_g(1)$.

3. THEOREM 5.3. *If $R_1 = e^{-H_1}$ and $R_2 = e^{-H_2}$ satisfy the inequality $R_1 J \ge R_2 J$, then $H_1 J \le H_2 J$.*

Indeed, $R_1^{1/2}J \ge R_2^{1/2}J$ by Theorem 4.4, and since $R_1^{1/2}$ and $R_2^{1/2}$ are moduli in turn, the same theorem gives us that $R_1^{1/4}J \ge R_2^{1/4}J$. Continuing in this way, we get that $e^{(-1/2^n)H_1}J \ge e^{(-1/2^n)H_2}J$, which implies that

$$2^n(J - e^{(-1/2^n)H_1}J) \le 2^n(J - e^{(-1/2^n)H_2}J).$$

The required inequality $H_1 J \le H_2 J$ is obtained by passing to the limit as $n \to \infty$.

In conclusion we remark that the last theorem and Theorem 4.4 on monotonicity of the logarithm and the square root are weak repercussions of the remarkable paper [2] of Löwner.

BIBLIOGRAPHY

1. A. I. Mal'tsev, *Foundations of linear algebra*, OGIZ, Moscow, 1948; English transl. of 2nd ed., Freeman, San Francisco, Calif., 1963.

2. Karl Löwner, *Über monotone Matrixfunktionen*, Math. Z. **38** (1933/34), 177–216.

Translated by H. H. MCFADEN

Amer. Math. Soc. Transl.
(2) Vol. **138**, 1988

A Theorem on the Modulus. II

UDC 517.5

V. P. POTAPOV

§6. A theorem on a product of moduli([1])

In connection with the multiplicativity of the class W of J-nonexpansive matrices the question inevitably arises of studying products composed of moduli $w_k = \prod_1^k r_i$ or $w_k = \prod_1^k e^{-H_i}$. In considering infinite sequences of such products we encounter two situations. The first, which is simpler, consists in having $w_{k+1} = w_k r_{k+1}$ for $k = 1, 2, \ldots$; in this case the question of the convergence of the sequence w_k arises naturally. The second and more general situation assumes that in passing from w_k to w_{k+1} both the factors themselves and their number change in an arbitrary way. Here criteria for the compactness of a family of matrices are essential.

In order to get oriented in the difficulties which arise, we dwell in a little more detail on the first case. Because multiplication is not commutative, a product of moduli is not a modulus in general; hence $w_k = R_k U_k$, where $U_k \neq I$, the J-forms $J - w_k J w_k^* = J - R_k^2 J$ (the defining structure of our theory!) are monotonically increasing, and their boundedness implies the existence of a limit.

The inequality $0 \leq J - R_k J \leq J - R_k^2 J$ and Theorem 4.4([1]) on the monotonicity of the modulus lead to the existence of the limit of the first component R_k of the polar representation, but it is difficult to say how the second component U_k behaves, especially if we take into account that in the case of an indefinite metric the group of J-unitary matrices is not compact.([2])

1980 *Mathematics Subject Classification* (1985 *Revision*). Primary 15A60; Secondary 47H09.
Translation of Teor. Funktsiĭ, Funktsional. Anal. i Prilozhen. Vyp. 39 (1983), 95–106;
MR **85e**:47054.

([1])For Part I, §§1–5, see the immediately preceding paper in this volume.

([2])Example:

$$J = \begin{pmatrix} -1 & 0 \\ 0 & 1 \end{pmatrix}, \qquad U = \begin{pmatrix} \cosh\theta & \sinh\theta \\ \sinh\theta & \cosh\theta \end{pmatrix}.$$

Our problem, thus, must be to "tame" U_k, presuming that even if a product of moduli is not a modulus, it nevertheless will not differ too much from a modulus. Moreover, we should take care to "prop up" the moduli R_k from below, since without this it is difficult to expect U_k to be bounded.

In considering the question of convergence or compactness of such products it is expedient to use the language of infinitesimal calculus, writing w_k in the form of a multiplicative integral

$$w_k = \int_0^k e^{-H(x)\,dx} = R_k U_k$$

where $H(x)$ is a step function equal to H_j on $[j-1, j]$. Generalizing the situation, we consider both the individual integral

$$w(t) = \int_0^t e^{-H(x)\,dx} \quad (0 \le t < +\infty)$$

and a collection of such integrals

$$w_\alpha(t) = \int_0^t e^{-H_\alpha(x)\,dx} \quad (0 \le t < +\infty),$$

where $H(x)$ and $H_\alpha(x)$ are arbitrary J-Hermitian-nonnegative matrix-valued functions integrable on each finite interval $[0, l]$:

$$\int_0^l \|H_\alpha(x)\|\,dx < +\infty.$$

Note first of all that $w(t)$ is invertible for all t, since along with

$$w(t) = \int_0^t e^{-H(x)\,dx}$$

the multiplicative integral

$$\int_0^t e^{H(x)\,dx} = w^{-1}(t)$$

also exists. We know that $w(t)$ and $w^{-1}(t)$ are absolutely continuous on any interval $[0, l]$; their derivatives are equal almost everywhere to $dw/dt = -w(t)H(t)$ and $dw^{-1}/dt = H(t)w^{-1}(t)$. It follows from the first equality that

$$J - w(t)Jw^*(t) = 2\int_0^t w(x)H(x)Jw^*(x)\,dx \ge 0.$$

Being an invertible J-nonexpansive matrix, $w(t)$ admits a polar representation: $w(t) = R(t)U(t)$. Since $J - w(t)Jw^*(t) = J - R^2(t)J$ and $w^{-1*}(t)Jw^{-1}(t) - J = JR^{-2}(t) - J$, the matrix-valued functions $R^2(t)$ and $R^{-2}(t)$ are absolutely continuous on any finite interval $[0, l]$. To see that $R^{-1}(t)$ and $R(t)$ have the same property, we establish for any $x_1, x_2 \in [0, l]$ the following inequalities:

$$\|R^{-1}(x_1) - R^{-1}(x_2)\| \le K_l \|R^2(x_1) - R^2(x_2)\| \tag{6.1.a}$$

$$\|R(x_1) - R(x_2)\| \le K_l \|R^{-2}(x_1) - R^{-2}(x_2)\| \tag{6.1.b}$$

Indeed, it follows from the integral representation (4.1) that

$$\|R^{-1}(x_1) - R^{-1}(x_2)\|$$

$$\le \|R^2(x_1) - R^2(x_2)\| \frac{2}{\pi} \int_0^{+\infty} \|[R^2(x_1) + t^2 I]^{-1}\| \, \|[R^2(x_2) + t^2 I]^{-1}\| t^2 \, dt$$

and it remains to estimate the integral after breaking it up into terms:

$$\int_0^{+\infty} = \int_0^N + \int_N^{+\infty}$$

The second term is written in the form

$$\int_N^{+\infty} \|[I + Q_t(x_1)]^{-1}\| \cdot \|[I + Q_t(x_2)]^{-1}\| \frac{dt}{t^2}, \qquad Q_t(x) = R^2(x)/t^2.$$

But the function $R^2(x)$, being continuous on $[0, l]$, is bounded, $\|R^2(x)\| \le M_1$, and, given a fixed q $(0 < q < 1)$, there is an N such that

$$\|Q_t(x)\| = \frac{1}{t^2} \|R^2(x)\| \le \frac{1}{t^2} M_1 \le q$$

for $t \ge N$ and for all $x \in [0, l]$. Then it is obvious that

$$\int_N^{+\infty} \le \frac{1}{1-q} \cdot \frac{1}{1-q} \cdot \frac{1}{N} = K_1.$$

To estimate the first integral we consider the function

$$R^2(x) + t^2 I = F(u)$$

of a point $u = (x, t)$ in the rectangle $T = \{(x, t): 0 \le x \le l, \ 0 \le t < N\}$. The function $F(u)$ is continuous everywhere in T, and $F^{-1}(u)$ exists, since $\sigma(F(u)) > 0$. But then $F^{-1}(u)$ is continuous on T and, consequently, is bounded, $\|F^{-1}(u)\| \le M_2$, which implies that $\int_0^N \le M_1 \cdot M_2 \cdot N_3/3 = K_2$.

Taking $K = (2/\pi)(K_1 + K_2)$, we get (6.1.a). Inequality (6.1.b) is proved similarly.

Finally, we conclude from the relations $U(t) = R^{-1}(t)w(t)$ and $U^{-1}(t) = w^{-1}(t)R(t)$ that the matrix-valued functions $U(t)$ and $U^{-1}(t)$ are also absolutely continuous.

Summing up, we conclude that the matrix-valued functions $w(t)$, $w^{-1}(t)$, $R^2(t)$, $R^{-2}(t)$, $R(t)$ and $R^{-1}(t)$ have derivatives almost everywhere that are integrable on any finite interval $[0, l]$.

Moreover, since the J-form $J - w(t)Jw^*(t) = J - R^2(t)J$ is monotonically increasing with increasing t, the preceding section gives us that the matrix-valued functions $R^2(t)J$ and $R(t)J$ are monotonically decreasing, while $R^{-2}(t)J$ and $R^{-1}(t)J$ are monotonically increasing.

MAIN THEOREM. *Let $W = \{w_\alpha\}$ be a set of J-nonexpansive matrices w_α that are discrete or continuous products of moduli written in the form*

$$w_\alpha = \int_0^{l_\alpha} e^{-H_\alpha(x)\,dx}, \qquad H_\alpha(x)J \geq 0,$$

where $0 \leq l_\alpha \leq +\infty$ and $\int_0^{l_\alpha} \|H_\alpha(x)\|\,dx < +\infty$, and let each of the matrices w_α satisfy the inequalities

$$\|J - w_\alpha J w_\alpha^*\| \leq C_1 \tag{6.2}$$

and

$$\|w_\alpha^{*-1} J w_\alpha^{-1} - J\| \leq C_2 \tag{6.3}$$

where C_1 and C_2 are constants. Let $w_\alpha = R_\alpha U_\alpha$ be the polar representation of w_α.

Under these conditions there exist constants $\kappa_1 = \varphi(C_1, C_2)$, $\kappa_2 = \psi(C_1, C_2)$, and $\kappa_3 = \chi(C_1, C_2)$, depending only on C_1 and C_2 but not on the concrete matrix $w_\alpha \in W$, such that $\|U_\alpha\| \leq \kappa_1$, $\int_0^{l_\alpha} \|H_\alpha(x)\|\,dx \leq \kappa_2$ and $\|w_\alpha\| \leq \kappa_3$.

PROOF. With each matrix

$$w_\alpha = \int_0^{l_\alpha} e^{-H_\alpha(x)\,dx}$$

in W we associate the multiplicative integral with variable upper limit

$$w_\alpha(t) = \int_0^t e^{-H_\alpha(x)\,dx} \qquad (0 \leq t \leq l_\alpha).$$

Let $w_\alpha(t) = R_\alpha(t)U_\alpha(t)$ be its polar representation. Theorems 4.3 and 4.4 and the conditions of the present theorem imply the inequalities

$$0 \leq J - R_\alpha(t)J \leq J - R_\alpha J \leq J - R_\alpha^2 J = J - w_\alpha J w_\alpha^* \leq C_1 I,$$
$$0 \leq JR_\alpha^{-1}(t) - J \leq JR_\alpha^{-1} - J \leq JR_\alpha^{-2} - J = w_\alpha^{*-1} J w_\alpha^{-1} - J \leq C_2 I,$$

which indicate the collective boundedness of the matrix-valued functions $R_\alpha(t)$ and $R_\alpha^{-1}(t)$: $\|R_\alpha(t)\| \leq C_1 + 1$, $\|R_\alpha^{-1}(t)\| \leq C_2 + 1$.

We prove that the matrix-valued functions $U_\alpha(t)$ and $U_\alpha^{-1}(t)$ can be expressed in terms of $R_\alpha(t)$ and $R_\alpha^{-1}(t)$ as follows:

$$U_\alpha(t) = \int_0^t e^{K_\alpha(x)\,dx}, \tag{6.4}$$

where

$$K_\alpha(x) = \tfrac{1}{2}[R_\alpha'(x)R_\alpha^{-1}(x) - R_\alpha^{-1}(x)R_\alpha'(x)], \tag{6.5}$$

$$H_\alpha(x) = -\tfrac{1}{2}U_\alpha^{-1}[R_\alpha^{-1}R_\alpha' + R_\alpha'R_\alpha^{-1}]U_\alpha. \tag{6.6}$$

After this, inequalities (6.2) and (6.3) enable us to get upper estimates of $U_\alpha(t)$ and $\int_0^{l_\alpha} \|H_\alpha(x)\|\, dx$. Omitting the index α for simplicity, we consider

$$w(t) = \int_0^t e^{-H(x)\,dx} = R(t)U(t) \qquad (0 \le t \le l).$$

We know that $dw/dt = -wH(t)$. This equality implies first of all that $R'U + RU' = -RUH$ or

$$R^{-1}R' + U'U^{-1} = -UHU^{-1} \tag{6.7}$$

On the other hand, we get successively that

$$[wJw^*]' = -2wHJw^*, \qquad [R^2J]' = -2RUHJU^*R^*,$$
$$R'R + RR' = -2RUHJU^*R^*J = -2RUHU^{-1}R,$$

which gives us that

$$R^{-1}R' + R'R^{-1} = -2UHU^{-1}. \tag{6.8}$$

Eliminating UHU^{-1} from (6.7) and (6.8), we find that $U'U^{-1} = K$, where $K = \frac{1}{2}[R'R^{-1} - R^{-1}R']$, which proves (6.5). Knowing U, we find H from (6.8). By (6.4) and the known estimate

$$\|U_\alpha^{\pm 1}\| \le \exp\left\{ \int_0^t \|K_\alpha(x)\|\, dx \right\} \le \exp\left\{ \int_0^t \|R^{-1}(x)\|\,\|R'(x)\|\, dx \right\}$$

for the multiplicative integral, and since $\|R^{-1}(x)\| \le C_2 + 1$, it follows that

$$\|U(t)^{\pm 1}\| \le \exp\left\{ (C_2 + 1) \int_0^l \|R'(x)\|\, dx \right\},$$

and it remains to estimate the integral $\int_0^l \|R'(x)\|\, dx$.

By what has been proved, the matrix-valued function $J - R(x)J$ is monotonically increasing as x increases; therefore, the derivative $\{J - R(x)J\}'$ is a nonnegative Hermitian matrix. The norm of such a matrix (the largest eigenvalue) does not exceed its trace (the sum of all the eigenvalues), which is equal to the sum of the diagonal elements. Therefore,

$$\int_0^l \|R'(x)\|\, dx = \int_0^l \|\{J - R(x)J\}'\, dx\|$$
$$\le \int_0^l \operatorname{tr}\{J - R(x)J\}'\, dx = \int_0^l \{\operatorname{tr}[J - R(x)J]'\}\, dx$$
$$= \operatorname{tr}[J - R(l)J] - \operatorname{tr}[J - R(0)J] = \operatorname{tr}[J - RJ]$$
$$\le m\|J - RJ\| \le m\|J - wJw^*\| \le mC_1.$$

Thus

$$\|U^{\pm 1}\| \le e^{mC_1(C_2+1)} = \varphi(C_1, C_2).$$

After this, it follows from (6.6) that

$$\|H(x)\| \le \|U^{-1}(x)\| \cdot \|U(x)\| \cdot \|R^{-1}(x)\| \cdot \|R'(x)\|$$

and hence

$$\int_0^l \|H(x)\|\, dx \le e^{2mC_1(C_2+1)}(C_2+1)mC_1 = \psi(C_1, C_2).$$

Finally, the polar representation gives us that

$$\|w\| \le \|R\|\, \|U\| \le (C_1+1)e^{mC_1(C_2+1)} = \chi(C_1, C_2).$$

The theorem is proved.

§7. Corollaries to the main theorem

We dwell on certain important facts following from the main theorem.

Let us note first of all that in a number of cases it is more convenient to use other equivalent conditions in place of (6.2) and (6.3).

1°. The conditions (6.2) and (6.3) of the main theorem are equivalent to

$$\|R_\alpha\| \le a_1 \qquad (7.1), \qquad \|R_\alpha^{-1}\| \le a_2 \qquad\qquad (7.2)$$

where a_1 and a_2 are constants.

Indeed, if (6.2) and (6.3) hold, then the inequalities

$$0 \le J - R_\alpha J \le J - w_\alpha J w_\alpha^*, \qquad 0 \le R_\alpha^{-1} J - J \le w_\alpha^{*-1} J w_\alpha^{-1} - J$$

imply $\|R_\alpha\| \le C_1 + 1$ and $\|R_\alpha^{-1}\| \le C_2 + 1$. Conversely, (7.1) and (7.2) imply the inequalities

$$\|J - w_\alpha J w_\alpha^*\| \le 1 + a_1^2 \quad \text{and} \quad \|w_\alpha^{*-1} J w_\alpha^{-1} J\| \le a_2^2 + 1.$$

After this, the first assertion of the main theorem can be reformulated as follows.

If R_α and R_α^{-1} are collectively bounded, then the J-unitary matrices U_α of the polar representation $w_\alpha = R_\alpha U_\alpha$ of the product of moduli

$$w_\alpha = \int_0^{l_\alpha} e^{-H_\alpha(x)\, dx}, \qquad H_\alpha(x)J \ge 0$$

have the same property.

However, a large role in applications is played by the second assertion of the theorem on the boundedness of the integrals $\int_0^{l_\alpha} \|H_\alpha(x)\|\, dx \le \kappa_2 = \psi(C_1, C_2)$.

2°. The conditions (6.2) and (6.3) of the main theorem are equivalent to the conditions

$$\|J - w_\alpha J w_\alpha^*\| \le C_1 \qquad\qquad (7.3)$$

and

$$|\operatorname{Det} w_\alpha| \ge \delta > 0 \qquad\qquad (7.4)$$

Indeed, suppose that (6.2) and (6.3) hold. We show that (7.4) holds:

$$\frac{1}{|\operatorname{Det} w_\alpha|^2} = \frac{1}{\operatorname{Det}(w_\alpha w_\alpha^*)} = \operatorname{Det}(w_\alpha^{*-1} \cdot w_\alpha^{-1})$$

$$= \operatorname{Det}(R_\alpha^{*-1} \cdot R_\alpha^{-1}) \le \|R_\alpha^{-1}\|^{2m} \le (C_2+1)^{2m} = \frac{1}{\delta^2}$$

and so $|\operatorname{Det} w_\alpha| \geq \delta > 0$. On the other hand, if (7.1) and (7.2) hold, then

$$\|R_\alpha\| \leq C_1 + 1, \quad |\operatorname{Det} R_\alpha| \geq \delta,$$

and, as follows from the structure of the inverse matrix, all the elements of R_α^{-1}, hence also $\|R_\alpha^{-1}\|$, are collectively bounded.

3°. The conditions (6.2) and (6.3) of the main theorem are equivalent to

$$\|J - w_\alpha J w_\alpha^*\| \leq C_1 \qquad (7.5); \qquad \int_0^l \operatorname{tr} H_\alpha(x)\, dx \leq K. \qquad (7.6)$$

It suffices to establish the equivalence of (7.4) and (7.6), and this follows from the equality

$$\operatorname{Det} w_\alpha = \exp\left(-\int_0^{l_\alpha} \operatorname{tr} H_\alpha(x)\, dx\right)$$

for $K = \ln 1/\delta$.

4°. THEOREM 7.1. *Let $H(t)$ be a J-Hermitian-nonnegative locally integrable matrix, and let $w(t) = \int_0^t e^{-H(x)\, dx}$, $t \geq 0$. For $H(x)$ to be integrable on the whole half-line, $\int_0^{+\infty} \|H(x)\|\, dx < +\infty$, it is necessary and sufficient that the following two conditions hold:*

$$\int_0^{+\infty} w(x)H(x)Jw^*(x)\, dx < +\infty, \qquad (7.7)$$

$$\sup \int_0^t \operatorname{tr} H(x)\, dx < +\infty. \qquad (7.8)$$

REMARK. The concretization (7.7), (7.8) of conditions (6.2) and (6.3) of the main theorem and the replacement of the family $\{H_\alpha(t)\}$ by a single matrix-valued function $H(t)$ are natural from the point of view of the theory of canonical systems of differential equations. The matrix-valued function $w(t)$ is the Wronskian of the system $dy/dt = -yH(t)$, and condition (7.7) is equivalent to the square integrability of all the solutions $y(t)$ with respect to the matrix weight $H(x)J$.

PROOF. The necessity is trivial, since, if $h = \int_0^{+\infty} \|H(x)\|\, dx < +\infty$, then the inequality

$$\|w(t)\| \leq \exp\left(\int_0^t \|H(x)\|\, dx\right) \leq e^h$$

implies that

$$\left\|\int_0^t w(x)H(x)Jw^*(x)\, dx\right\| \leq e^{2h}\int_0^t \|H(x)\|\, dx \leq he^{2h}.$$

But since

$$2\int_0^t w(x)H(x)Jw^*(x)\, dx = J - w(t)Jw^*(t),$$

it follows that the left-hand is monotonically increasing as t increases, and, being bounded, it has a limit as $t \to +\infty$; that is, (7.7) holds. Further, the convergence of the integral $\int_0^\infty \|H(x)\|\, dx$ implies the convergence of $\int_0^\infty H(x)\, dx$, and hence the convergence of $\int_0^{+\infty} \operatorname{tr} H(x)\, dx$.

The sufficiency of conditions (7.7) and (7.8) follows from the second assertion of the main theorem.

The most intuitive illustration of the main theorem is perhaps the following proposition, mentioned earlier.

5°. THEOREM 7.2. *The infinite product of moduli*

$$\overset{\infty}{\underset{1}{\overset{\curvearrowright}{\prod}}} e^{-H_j}, \qquad H_j J \geq 0$$

converges if and only if $\sum_1^\infty H_j$ *converges.*

To prove the necessity we note that the sequences

$$w_n = \overset{n}{\underset{1}{\overset{\curvearrowright}{\prod}}} e^{-H_j} \quad \text{and} \quad w_n^{-1} = \overset{n}{\underset{1}{\overset{\curvearrowleft}{\prod}}} e^{H_j}$$

have limits and, consequently, are bounded. But then the corresponding J-forms are also bounded; that is, conditions (6.2) and (6.3) of the main theorem are satisfied. By the second assertion, $\sum_1^n \|H_j\| \leq \kappa_2 = \psi(C_1, C_2)$, which implies that $\sum_1^\infty \|H_j\| < +\infty$.

The proof of sufficiency has already been discussed. We mention only that in the case of matrices of finite order the convergence of $\sum_1^\infty \|H_j\| < +\infty$ implies successively the convergence of $\sum_1^\infty H_j J$, $\sum_1^\infty \operatorname{tr} H_j J$ and $\sum_1^\infty \|H_j\|$, and conversely.

6°. Recall that for any matrix-valued function $H(t)$ integrable on $[0, l]$ the multiplicative Lebesgue integral

$$\overset{t}{\underset{0}{\overset{\curvearrowright}{\int}}} e^{-H(t)\, dt}$$

is defined as a multiplicative Stieltjes integral:

$$\overset{t}{\underset{0}{\overset{\curvearrowright}{\int}}} e^{-H(t)\, dt} = \overset{l}{\underset{0}{\overset{\curvearrowright}{\int}}} e^{-d\Sigma(t)}, \quad \text{where} \quad \Sigma(t) = \int_0^t H(x)\, dx;$$

$\Sigma(t)$ is an absolutely continuous matrix-valued function, and $d\Sigma(t)/dt = H(t)$ almost everywhere. In the situation being considered here, $H(t)J \geq 0$, and hence $\Sigma(t)$ is a monotonically J-nondecreasing matrix-valued function. We use this circumstance to unify the choice of the variable of integration. With this purpose we consider the trace $\tau(t) = \sum_{j=1}^m \sigma_{ij}(t) = \operatorname{tr} \Sigma(t) J$ of the matrix $\Sigma(t) J$ and introduce the new variable θ, setting $\theta = \tau(t)$. Let $[0, L]$ be the interval onto which the absolutely continuous monotonically nondecreasing function $\theta = \tau(t)$ maps the interval $[0, l]$. Despite the fact that the function $\theta = \tau(t)$ can have intervals of constancy, so that to a single value θ_0 there correspond infinitely many values of $t \in [a, b]$, the matrix-valued function $\Sigma(t)$ depends in a single-valued fashion on θ. Indeed, we consider two values of the argument $t : t_1 < t_2$, corresponding to a single given value $\theta = \theta_0$, and the corresponding values $\Sigma(t_1)$ and $\Sigma(t_2)$ of the matrix-valued function $\Sigma(t)$. Since

$$\|\Sigma(t_2) - \Sigma(t_1)\| = \|[\Sigma(t_2) - \Sigma(t_1)]J\|$$
$$\leq \operatorname{tr}[\Sigma(t_2) - \Sigma(t_1)]J = \tau(t_2) - \tau(t_1) = \theta_0 - \theta_0,$$

it follows that $\Sigma(t_2) = \Sigma(t_1)$. We define the matrix-valued function $E(\theta)$ by setting $E(\theta) = \Sigma(t)$, $\theta = \tau(t)$. Obviously,

$$\operatorname{tr} E(\theta)J = \operatorname{tr} \Sigma(t)J = \tau(t) = \theta$$

which implies that

$$\|E(\theta_2) - E(\theta_1)\| = \|[E(\theta_2) - E(\theta_1)]J\|$$
$$\leq \operatorname{tr}[E(\theta_2) - E(\theta_1)]J = \theta_2 - \theta_1,$$

for any $\theta_1, \theta_2 \in [0, L]$, $\theta_1 < \theta_2$. Thus, $E(\theta)$ satisfies the Lipschitz condition $\|E(\theta_2) - E(\theta_1)\| \leq \theta_2 - \theta_1$.

But then, first, the multiplicative Stieltjes integral

$$\int_0^L e^{-dE(\theta)},$$

exists, and, second, $E(\theta)$ is absolutely continuous and thus has almost everywhere an integrable derivative $M(\theta) = dE/d\theta$, $M(\theta)J \geq 0$, such that $E(\theta) = \int_0^\theta M(x)\,dx$ and such that $\operatorname{tr} M(\theta)J = 1$ almost everywhere. By the foregoing, the multiplicative Lebesgue integral

$$\int_0^L e^{-M(\theta)\,d\theta} = \int_0^L e^{-dE(\theta)}$$

exists. Finally, we have

$$\int_0^L e^{-M(\theta)\,d\theta} = \int_0^l e^{-H(t)\,dt}$$

which is equivalent to the obvious equality

$$\int_0^L e^{-dE(\theta)} = \int_0^L e^{-d\Sigma(t)}.$$

Writing a product of moduli below in the form of a multiplicative Lebesgue integral

$$\int_0^l e^{-H(t)\,dt}, \qquad H(t)J \geq 0,$$

we assume that the condition $\operatorname{tr} \int_0^t H(x)J\,dx = t$ holds.

An important role in compactness questions is played by

THEOREM 7.3. *If the family $W = \{w_\alpha\}$ of products of moduli satisfies the conditions of the main theorem and the w_α are written in the form*

$$w_\alpha = \int_0^{\curvearrowleft l_\alpha} e^{-H_\alpha(t)\,dt}, \qquad H_\alpha(t)J \geq 0, \quad \operatorname{tr} H_\alpha(t)J = 1 \text{ a.e.,}$$

then the upper limits l_α of the multiplicative integrals are bounded, and each sequence w_{α_k} has a subsequence $w_{\alpha_{k_\nu}}$ converging a.e. to a matrix

$$w = \int_0^{\curvearrowleft l} e^{-H(t)\,dt}, \qquad H(t)J \geq 0, \quad \operatorname{tr} H(t)J = 1.$$

PROOF. The boundedness of $\{l_\alpha\}$ follows from the second assertion of the main theorem and the inequality $\operatorname{tr} H_\alpha(t)J \leq m\|H_\alpha(t)J\| = m\|H_\alpha(t)\|$, where m is the order of the matrix w_α. Indeed,

$$l_\alpha = \int_0^{l_\alpha} 1\,dt = \int_0^{l_\alpha} \operatorname{tr} H_\alpha(t)J\,dt \leq m \int_0^{l_\alpha} \|H_\alpha(t)\|\,dt = m\psi(C_1, C_2).$$

Using the second assertion of the main theorem, we select from an arbitrary sequence in the family $\{w_\alpha\}$ of matrices a subsequence w_{α_k} such that $l_{\alpha_k} \to l$. The interval $[0, l]$ can be called the "Procrustean bed", having in mind the procedures to which the matrices w_{α_k} must be subjected.

If $l_{\alpha_k} > l$, then $\Sigma_{\alpha_k}(t)$ is truncated to $\tilde{\Sigma}_{\alpha_k}(t)$, which is defined on $[0, l]$ and coincides there with $\Sigma_{\alpha_k}(t)$. It is clear that the sequences $\tilde{w}_{\alpha_k}$ and w_{α_k} are cofinal,$(^3)$ since

$$\|w_{\alpha_k} - \tilde{w}_{\alpha_k}\| = \left\| \tilde{w}_{\alpha_k} \int_l^{\curvearrowleft l_{\alpha_k}} e^{-H_{\alpha_k}(t)\,dt} - \tilde{w}_{\alpha_k} \right\|$$

$$\leq \|\tilde{w}_{\alpha_k}\| \left\| \int_l^{\curvearrowleft l_{\alpha_k}} e^{-H_{\alpha_k}(t)\,dt} - I \right\|$$

$$\leq \exp\left(\int_0^l \|H_{\alpha_k}(t)\,dt \right) \left\| \int_l^{l_{\alpha_k}} H_{\alpha_k}(t)\,dt \right\| \exp\left(\int_l^{l_{\alpha_k}} \|H_{\alpha_k}(t)\|\,dt \right)$$

$$\leq e^{\psi(C_1, C_2)}(l_{\alpha_k} - l) \to 0.$$

But if $l_{\alpha_k} < l$, then $\Sigma_{\alpha_k}(t)$ is stretched out to the interval $[0, l]$ by the equality $\tilde{\Sigma}_{\alpha_k}(t) = \Sigma_{\alpha_k}(l_{\alpha_k})$ $(l_{\alpha_k} \leq t \leq l)$. It is clear that here $\tilde{w}_{\alpha_k} = w_{\alpha_k}$.

Finally, if by chance $l_{\alpha_k} = l$, then $\Sigma_{\alpha_k}(t)$ remains unharmed: $\tilde{w}_{\alpha_k} = w_{\alpha_k}$.

It is easy to see that the cofinal sequence $\tilde{w}_{\alpha_k}$ constructed in this way has the property that

$$\tilde{w}_{\alpha_k} = \int_0^{\curvearrowleft l} e^{-d\tilde{\Sigma}_{\alpha_k}(t)}$$

$(^3)$Recall that sequences a_n and b_n are said to be *cofinal* if $a_n - b_n \to 0$.

where $\operatorname{tr} \tilde{\Sigma}_{\alpha_k}(t) \leq t$, and the strict inequality holds for $l_{\alpha_k} < l$ on an infinitesimally small interval $[l_{\alpha_k}, l]$.

The Helly selection theorem comes to be used next. Since $\tilde{\Sigma}_{\alpha_k}(t)$ is monotonically nondecreasing, its diagonal elements $\sigma_{jj}^{(k)}(t)$ have the same property, while the other elements (in view of the nonnegativity of the principal minors of second order) satisfy the inequality

$$|\Delta\sigma_{ij}^{(k)}(t)| \leq \sqrt{\Delta\sigma_{ii}^{(k)}(t)} \cdot \sqrt{\Delta\sigma_{jj}^{(k)}(t)} \leq \tfrac{1}{2}[\Delta\sigma_{ii}^{(k)}(t) + \Delta\sigma_{jj}^{(k)}(t)]$$

and are thereby functions of bounded variation.

These facts enable us to select from $\tilde{\Sigma}_{\alpha_k}(t)J$ a subsequence $\Sigma_{\alpha_{k_p}}(t)J$ converging at each point of $[0, l]$ to a monotone matrix-valued function $\Sigma(t)J$ such that $\operatorname{tr}\Sigma(t)J = t$. Letting $d\Sigma/dt = H(t)$, we consider the matrix

$$w = \int_0^l \curvearrowleft e^{-H(t)\,dt}.$$

To prove that $w_{\alpha_{k_p}} \to w$, we consider a partition of the interval $[0, l]: 0 = t_0 < t_1 < t_2 < \cdots < t_n = l$. Representing the multiplicative Stieltjes integrals as products

$$\int_0^l \curvearrowleft = \prod_{j=1}^n \curvearrowleft \cdot \int_{t_{j-1}}^{t_j} \curvearrowleft$$

we first use the estimate

$$\|w_{\alpha_{k_\nu}} - w\| \leq M \sum_{j=1}^n \left\| \int_{t_{j-1}}^{t_j} \curvearrowleft e^{-H_{\alpha_{k_\nu}}(t)\,dt} - \int_{t_{j-1}}^{t_j} \curvearrowleft e^{-H(t)\,dt} \right\|,$$

and then the estimates

$$\left\| \int_{t_{j-1}}^{t_j} \curvearrowleft e^{-\tilde{H}_{\alpha_{k_\nu}}(t)\,dt} \int_{t_{j-1}}^{t_j} \curvearrowleft e^{-H(t)\,dt} \right\|$$

$$\leq \left\| \tilde{\Sigma}_{\alpha_{k_\nu}}(t_j) - \Sigma(t_j) \right\| + \left\| \tilde{\Sigma}_{\alpha_{k_\nu}}(t_{j-1}) - \Sigma(t_{j-1}) \right\| + \|R_{\alpha_{k_\nu}}^{(j)}\| + \|R^{(j)}\|.$$

Choosing a fixed partition fine enough that

$$\sum_{j=1}^n (\|R_{\alpha_{k_\nu}}^{(j)}\| + \|R^{(j)}\|) < \frac{\varepsilon}{2},$$

we get for sufficiently large ν that $\|\tilde{w}_{\alpha_{k_\nu}} - w\| < \varepsilon$, i.e.,

$$\lim_{\nu\to\infty} \tilde{w}_{\alpha_{k_\nu}} = w,$$

and hence also $\lim_{\nu\to\infty} w_{\alpha_{k_\nu}} = w$, which proves Theorem 7.3.

Translated by H. H. McFADEN

ABCDEFGHIJ — 898